Surface Chemistry

Surface Chemistry

Elaine M. McCash

University of York

and

Sentec Ltd.

Cambridge

OXFORD

UNIVERSITY PRESS

OXFORD
UNIVERSITY PRESS

Great Clarendon Street, Oxford OX2 6DP

Oxford University Press is a department of the University of Oxford.
It furthers the University's objective of excellence in research, scholarship,
and education by publishing worldwide in

Oxford   New York

Athens Auckland Bangkok Bogotá Buenos Aires Calcutta
Cape Town Chennai Dar es Salaam Delhi Florence Hong Kong Istanbul
Karachi Kuala Lumpur Madrid Melbourne Mexico City Mumbai
Nairobi Paris São Paulo Singapore Taipei Tokyo Toronto Warsaw

with associated companies in   Berlin   Ibadan

Oxford is a registered trade mark of Oxford University Press
in the UK and in certain other countries

Published in the United States
by Oxford University Press Inc., New York

A catalogue record for this book is available from the British Library

Library of Congress Cataloging in Publication Data
(Data applied for)

ISBN 0 19 850328 8

Typeset by EXPO Holdings, Malaysia
Printed in Great Britain
on acid-free paper by T.J. International Ltd.,
Padstow, Cornwall

*for Helen
&
Matthew*

PREFACE

My good friend Colin Banwell leaned back in his chair and sipped his wine. He sighed contentedly. The manuscript for the fourth edition of *Fundamentals of Molecular Spectroscopy* had just been stowed away safely, ready for delivery to the publisher. 'You'll have to do one on your own now.' he said with a wicked glint in his eye. So here it is, seven years, two children, two academic promotions and a career change later. Colin knows that I can never resist a challenge.

Originally I set out to write a surface chemistry textbook to fill a gap that has made it difficult to find the right book to recommend to undergraduates. As I wrote it I realized that authors, selfishly, write for themselves. Whatever the motivation, however, I hope that what I have written gives insights into both the way we have tended to discuss and study surfaces in the past and the direction in which the field is moving. Surface science has come of age; there are now many techniques we can use and, as technologies advance, the theoretical models available continue to be developed to keep pace with them. It should be noted that this book is written for students, not their supervisors, and aims to give a real 'feel' for the subject, rather than a detailed, laborious understanding.

I am indebted to several people who have helped me in the preparation of the manuscript. First I must thank John Olive for turning my misshapen, hand-drawn sketches into diagrams of beauty and, more importantly, accuracy.

I am grateful to Professor Norman Sheppard FRS (Emeritus Professor, University of East Anglia) for his excellent and, as always, incisive comments on a significant portion of the manuscript; and to my former colleagues at York, Professor Jim Matthew and Dr Steve Tear (Department of Physics) and my husband Dr John Wilkes, for their views on the remainder. I am indebted to all those colleagues who provided me with ideas and figures which have proved invaluable. I must also thank my research group for their support and assistance, in particular, Drs Jemimah Eve, John Camplin and Jeanette Cook. I am greatly indebted to Dr Bill Allison, Andrew Jardine and Donald MacLaren in the Cavendish Laboratory, Cambridge, for some lively, illuminating discussions and for proofreading the final revisions during my highly enjoyable sabbatical term there.

I also thank Melissa Levitt, my commissioning editor from OUP, for talking me into writing the book in the first place and for her forbearance when my son's arrival disrupted the delivery of the manuscript. Finally, I am grateful to my husband and my daughter for their support and tolerance during the lengthy writing process.

Cambridge   E. McC.
November 2000

CONTENTS

4 Surface reactions and reactivity

5 Ultrathin films and interfaces

1

Introduction

1.1 Motivations

The interfaces between states of matter have been of intense interest to scientists for thousands of years. This book introduces solid surfaces and their physical and chemical behaviour.

Solid surfaces are of particular importance to everyday life. Some of the most important areas in which they play a vital role are summarized in Fig. 1.1. One of the most economically important areas is that of heterogeneous catalysis, where it is found that 90 per cent of the world's industrial output of materials such as fertilizers and plastics are produced from reactions on heterogeneous catalysts. Much effort has been and continues to be devoted to studies in this field; in fact, such are the advantages of surface reactions that when new homogeneous catalytic processes are developed, an

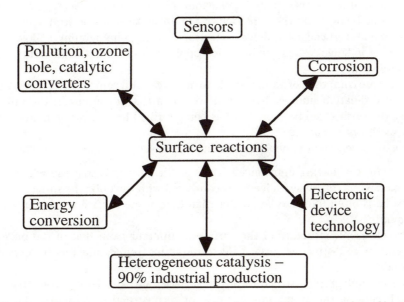

Fig. 1.1 Schematic diagram of the importance of surface reactions in a number of key areas.

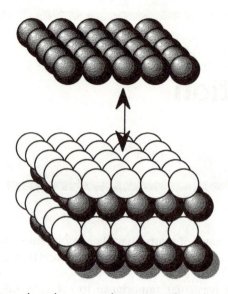

Fig. 1.2 The surface atoms have lower coordination numbers than the related bulk atoms. The surface is formed by effectively removing a plane of bulk atoms.

immense amount of work is usually carried out to transfer the reaction on to an appropriate surface. Another area of vast importance is that of corrosion, where surface reactions have a devastating effect. Surfaces and their interactions are important in the formation and behaviour of electronic devices, the action of lubricants (tribology) and the depletion of the ozone layer in the stratosphere. Their behaviour is also crucial in the fields of electrochemistry, photography, colloids, optics, data storage and, increasingly, in biological applications such as the use of membranes and biosensors.

Surfaces are composed of atoms which do not have a full complement of neighbours, i.e. the coordination number is much lower than that for atoms in the bulk solid because the surface can be thought of as being formed by 'removing' a layer of bulk atoms, as shown in Fig. 1.2.

There are several consequences of having this lower co-ordination number:

(a) The effect of having unbalanced forces at the surface is that the whole surface region is at a relatively high energy compared with the bulk; the surface energy of a solid is often found to be greater than that of the surface tension in a typical liquid.

(b) The electronic structure of the surface is different from that of the bulk solid. Metals have distinctive surface electronic states; semiconductors have 'dangling bonds'.

(c) The crystallographic structure of the surface can be different from that of the bulk because the atoms might move (or reconstruct) to minimize the surface energy.

(d) The binding or 'adsorption' of gases is strongly favoured at the surface, either by the formation of chemical bonds (chemisorption) or by weak van der Waals-type interactions causing physical adsorption (physisorption). Once adsorbed, the surface may enable the molecule to undergo reaction.

The study of surfaces and their behaviour requires a wide range of chemical knowledge and understanding; this book assumes a knowledge of the basic principles of physical chemistry, especially kinetics, dynamics and thermodynamics. A general knowledge of other important areas in chemistry is also useful; parallels can be drawn with, for example, polymers and organometallics (molecular fragments bound to surfaces often behave like ligands in organometallic complexes).

1.2 The single crystal surface

Many materials, including metals, semiconductors and insulators (such as NaCl), can be grown as single crystals of considerable size (typically ~10–15 mm or more in diameter). These can be cut along specific orientations to reveal surfaces with a high degree of order and simplicity; these single crystal surfaces offer a limited number of types of site on which molecules and molecular fragments can bind. This relative simplicity enables us to analyse how the molecule/molecular fragment adsorbs on and interacts with the surface.

The use of single-crystal surfaces does have its disadvantages as well. The number of atoms in the surface is small in comparison with the number of atoms found in the condensed phase (either liquid or solid), which means that surfaces are difficult to study experimentally. For example, if we consider a cube of water ice of side 1 cm, the volume (1 cm^3) contains about 10^{22} molecules; the six faces of the cube each contain about 10^{15} atoms cm^{-2}. The six to seven orders of magnitude difference obviously leads to stringent sensitivity requirements for studying the surface compared to the condensed phase.

Single crystals of metals can be prepared using a number of methods; for example, some metals can be grown by a seeding method where a rod of the single crystal is drawn slowly from a metal melt. Single crystals of semiconductors can be produced by vapour deposition. However it is formed, such a crystal can be cut to the desired orientation, revealing the required face. Mechanical polishing using a range of diamond pastes, and etching either chemically or electrochemically (using an etch designed specifically for the material), completes the preparation.

Once prepared, the crystal must be mounted in an ultrahigh vacuum (uhv) chamber. This is necessary because the second major disadvantage of using single crystal samples is that in order to attain atomic cleanness, it is necessary to carry out experiments in an environment which will minimize contamination. Hence uhv systems usually operate in the 1×10^{-10} mbar pressure range, which is about the same pressure as is found in space between the planets in our solar system.

1.3 Techniques for studying surfaces

Techniques which can be used to study surfaces have been developed rather more slowly than those for studying bulk samples and gas-phase phenomena. This is because of the small sample sizes involved and the inherent complexity of the surface systems. However, as technological advances have been made, it has proved possible to apply a whole range of techniques to surfaces, often with surprising results. The philosophy behind this book is to discuss the chemistry of surfaces by drawing on our current understanding, which is built on the conclusions of experimental and theoretical studies. For this reason the book is not 'driven' by the techniques which are used in surface science and so descriptions of the plethora of methods which are now available to study the surface are not given. Some of the most important techniques *are* described in detail, but only when they are needed for understanding and example. This section indicates a few examples of the types of technique which will be encountered.

The analysis of single crystal surfaces and atoms/molecules/molecular fragments which bind to them (termed adsorbates) has largely been developed on the basis of techniques used for bulk-solid, liquid and gas-phase samples. Most techniques therefore rely on the use of particle scattering, spectroscopy and diffraction. In addition, because of the special nature of the surface, several techniques have been developed which are surface specific, the most revolutionary example of this being scanning tunnelling microscopy (STM), which enables us to 'see' atoms. Photons, electrons, ions and atoms can all be used as probe species for investigating the surface, with the resulting photons, electrons, ions and atoms being detected as shown in Fig. 1.3.

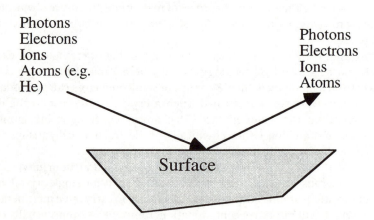

Fig. 1.3 Techniques applied to the study of surfaces involve the use of photons, electrons, ions and atoms as probe species.

A few examples of these techniques, all of which are described in more detail later in the book, are as follows:

(a) Auger electron spectroscopy, which is an example of an 'electron in, electron out' technique that provides a chemical fingerprint. The surface is bombarded with high-energy electrons, which remove electrons from the low-lying inner core shells of the atoms. An electron from a higher energy level will then drop down to fill the low-lying level, releasing energy which is taken up by an electron in another high-lying energy level. This 'Auger' electron is ejected with an energy characteristic of the three energy levels involved. Thus these 'Auger energies' are characteristic of the element.

(b) An example of 'photons in, photons out' is reflection–absorption infrared spectroscopy (RAIRS) which can be applied to adsorbates on surfaces. IR beams are used at grazing angles of incidence to the surface and, for adsorption on a metal surface, the vibrations detected are those with a dynamic dipole perpendicular to the surface. From an analysis of which vibrational modes are present or absent in the spectrum it is possible to determine the symmetry and orientation of the adsorbed molecule or molecular fragment. More complicated selection rules exist for adsorbates on semiconductor surfaces, but it is still possible to gain important insights into the nature of the adsorption process on these surfaces.

(c) Photoelectron spectroscopy (PES) has also found wide application to the study of surfaces, using either UV or X-ray photons for excitation. The photoelectrons emitted from the surface region provide us with information on the electronic structure of the surface and its chemical composition. The bonding of adsorbed species can also be investigated with PES.

(d) Diffraction phenomena can be observed from the surface using electrons or He atoms as probes. Diffraction is used to study surface structure and, to a certain extent, order on the surface.

(e) Several methods can be used to detect species that are evolved from the surface. For example, temperature programmed desorption (TPD) uses mass spectrometry to detect the molecules which are desorbed from the surface as it is heated rapidly through a given temperature range. This gives valuable information on the activation energy barriers to desorption and hence on the strength of bonding of the adsorbate to the surface. Temperature programmed reaction spectroscopy (TPRS) applies the same method to the detection of reaction products, giving valuable insights into the reactivity of the surface.

Surface techniques provide the key to understanding an extensive range of surface phenomena and this is illustrated by examples described in the text. The examples encompass topics such as carbon monoxide and hydrocarbon adsorption, heterogeneous reactions in the atmosphere and investigations of reaction dynamics.

Surface science is now a mature field and the techniques applied to surfaces have provided data which have given it a rigorous footing. We can now contemplate, and often answer, questions such as: What are the atoms on the surface? Why are they there? What are the bond energies involved? How does the adsorption/reaction proceed? As our understanding increases we can determine the factors which govern bond making and breaking and this will ultimately allow us to achieve the goal of influencing and controlling processes at the surface.

2
The clean surface

2.1 Ideal structures

Atoms in solids take up regular, ordered crystallographic structures which result from minimizing their free energy, G. It should be noted that the minimum value of G obtained is the equilibrium value, which depends on the precise pressure and temperature conditions. Thus the crystallographic structure adopted may change with pressure or temperature. For metals such as copper, platinum and nickel the 'usual' adopted structure is face-centred cubic (fcc) in which each atom has 12 nearest neighbours, while for metals such as iron and molybdenum a more open body-centred cubic (bcc) structure is usually adopted, where each of the atoms has eight nearest neighbours. The unit cells for these common structures are shown in Fig. 2.1. At this stage it is important to revise a certain amount of crystallography and to relate it to the parameters that are important in surface science.

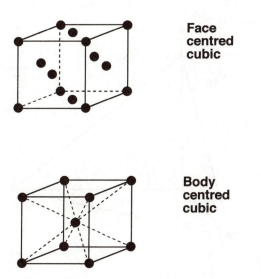

Face centred cubic

Body centred cubic

Fig. 2.1 Unit cells for (a) face centred cubic and (b) body centred cubic crystal structures.

2.1.1 Miller indices

Different crystallographic planes are exposed by slicing through single crystals at different angles and these are described, for simple cubic and fcc crystals, in terms of 'Miller indices'. The Miller index describes the vector which is perpendicular to the plane of interest and can be found by following some straightforward rules.

To find the Miller index

   (i) Find the intercepts of the plane with the three crystallographic directions or axes in terms of the primitive vectors ($\underline{a},\underline{b},\underline{c}$);
  (ii) Take the reciprocals;
 (iii) Multiply each by the smallest common number that makes each an integer, to produce the Miller index (h,k,l).

For example, to determine the Miller index of the plane sketched in Fig. 2.2.

(i) Intercepts	2, 5, 4
(ii) Reciprocals	$\frac{1}{2}, \frac{1}{5}, \frac{1}{4}$
(iii) Multiply by 20	10, 4, 5

So the Miller index is (10,4,5).

For hexagonal close-packed (hcp) crystals the Miller–Bravais indexing system is used. This takes into account the fact that a crystal has a hexagonal structure and so can be defined by four vectors, three of which are interdependent.

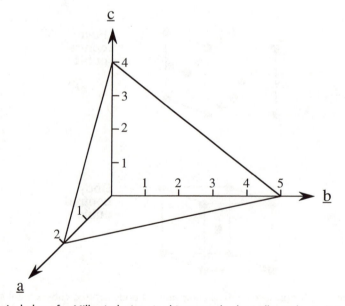

Fig. 2.2 A typical plane for Miller Indexing. In this example the Miller index is (10, 4, 5).

2.1.2 Low indexed faces

In studies of surfaces it is very common to investigate the low indexed faces as these have the most ordered structures and offer the smallest numbers of sites for adsorption. They are therefore thought of as providing the simplest systems for analysis. Some examples are shown in Fig. 2.3, where the commonest faces to be

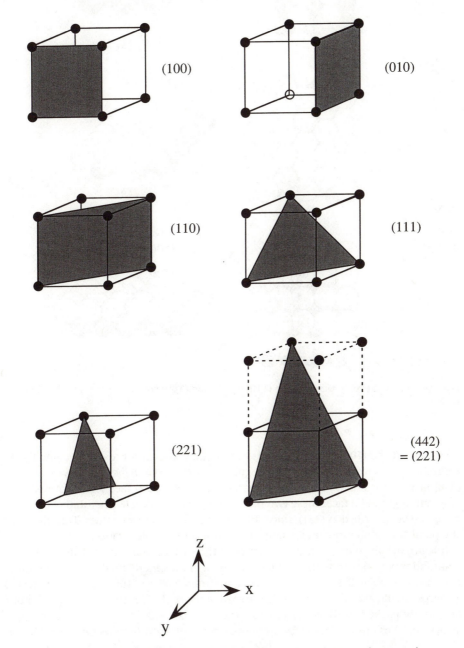

Fig. 2.3 The (100), (010), (110), (111) and (221) surfaces of a primitive cubic crystal.

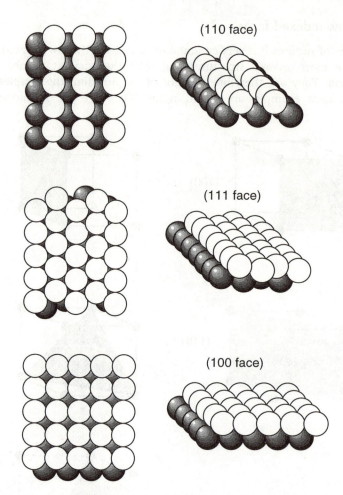

Fig. 2.4 Plan and profile views of the (110), (111) and (100) surfaces using a hard sphere model.

investigated, the (100), (110) and (111) surfaces, are shown for a simple cubic lattice. Figure 2.3 also shows that for a simple cubic lattice, the (100) face is equivalent to the (010) face and the (001) face. The construction of the (221) face is shown next to a diagram of the (442) face; these two planes are parallel to each other and (442) would normally be described as (221), since this has the lowest integer values. This illustrates the point that Miller indices actually describe sets of parallel planes.

It is important to have a mental picture of the surface under investigation and so a common way to illustrate the surface is to use a 'hard sphere' model, as shown for an fcc structure in Fig. 2.4, where both plan and profile views of the (110), (111) and (100) surfaces are illustrated. Important features to note are that for the (111) face the atoms are tightly packed in the surface, while those on the (100) face are in a more open structure. Those in the (110) face are more open still, with furrows between adjacent

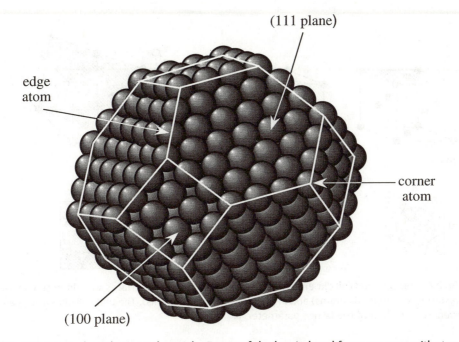

(111 plane)

edge
atom

corner
atom

(100 plane)

Fig. 2.5 A typical catalyst metal particle. Facets of the low indexed faces are seen with step edges and corner atoms which combine to minimise the surface energy.

top rows, as if in a ploughed field. The openness of the surface structure has important implications for surface energy and structure considerations.

It might be expected that the study of these low-indexed single-crystal faces would present a false comparison to more complex surfaces, such as are found in heterogeneous catalysts. However, the parallel is somewhat greater than might at first be predicted. In a typical catalyst the surface is usually composed of a metal oxide support material combined with an 'active' metal phase and often some kind of promoter. The metal tends to form particles, such as that shown in Fig. 2.5. In order to minimize the surface energy, terraces of the lowest energy faces, such as (111) and (100), tend to be exposed, in combination with step edges and corner atoms, as illustrated in the figure. Studies of adsorption on single crystal surfaces can therefore often be related directly to investigations of the real catalyst, where identical adsorption sites may be adopted. It should be noted that the ultimate form of the metal particles is also critically dependent on the interaction between the metal and the support material.

2.1.3 Surface atom densities

It is useful to make a quantitative estimate of N_s, the number of atoms per unit surface area, and this can be done from a knowledge of the crystal structure and the lattice parameter, a, which defines the length of the unit cell of the fcc lattice. It is first

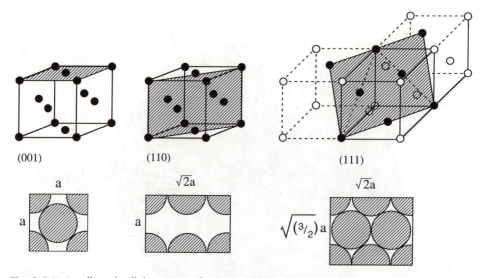

(001)   (110)   (111)

Fig. 2.6 Unit cells and cell dimensions for (001), (110) and (111) surfaces superimposed on the crystal lattice (upper diagrams) and in plan view (lower diagrams). The unit cell dimensions are given as a function of the lattice parameter, a.

necessary to calculate the area of the unit cell, bearing in mind the geometry of the face under consideration.

Figure 2.6 shows the unit cells for an fcc crystal with the (001), (110) and (111) faces superimposed. Beneath these diagrams are the plan views of these faces, showing the lengths of the sides in terms of a.

From Fig. 2.6 it can be seen that the (100) face contains two atoms per unit cell; this arises from the central atom in the cell plus four quarter shares of the four corner atoms. Each of the corner atoms is shared by four unit cells. It can be seen from the figure that the area of the (100) unit cell face is a^2. For a crystal where $a = 0.352$ nm, $a^2 = 1.24 \times 10^{-15}$ cm^2.

N_s is defined as the number of atoms in the unit cell, divided by the area of the unit cell. In the case of the (100) face then, $N_s = 2/a^2 = 2/(1.24 \times 10^{-15}) = 1.61 \times 10^{15}$ cm^{-2}.

Similarly for the (110) and (111) faces where there are two and four atoms per unit cell, respectively: $N_s = 2/\sqrt{2}a^2 = 1.14 \times 10^{15}$ cm^{-2} for (110) and $N_s = 4/\sqrt{3}a^2 = 1.86 \times 10^{15}$ cm^{-2} for (111) surfaces of the same crystal.

2.1.4 Unit meshes

The unit cell of a two-dimensional (2-d) surface structure is called a unit mesh and the corresponding 2-d lattice is called a surface net. If the unit mesh is translated across the surface, it can be built up to form a surface net. Some examples of unit meshes and surface nets for fcc and bcc surfaces are shown in Fig. 2.7.

Until now we have only considered ideal surfaces, which are well ordered and contain no defects. 'Real' single crystal surfaces have 'kinks', 'steps' and defects. These can be minimized by careful preparation of the single-crystal surface, but are rarely

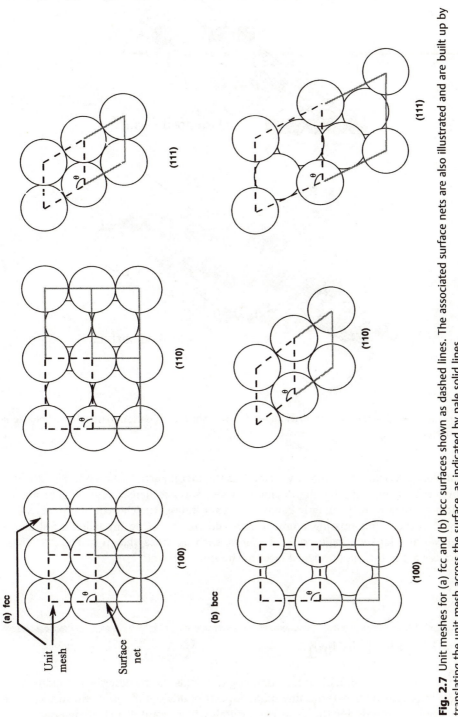

Fig. 2.7 Unit meshes for (a) fcc and (b) bcc surfaces shown as dashed lines. The associated surface nets are also illustrated and are built up by translating the unit mesh across the surface, as indicated by pale solid lines.

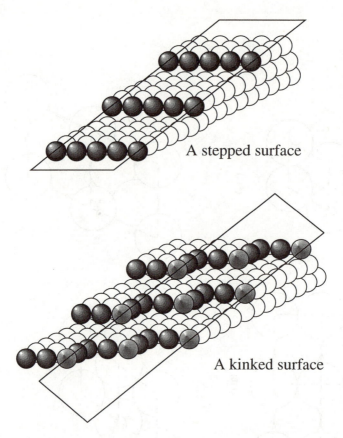

Fig. 2.8 Examples of stepped and a kinked surfaces which can be used to model the behaviour of defects.

completely absent. However, if we consider the metal particle shown in Fig. 2.5, it is obvious that we must be able to model these sorts of defect sites to gain a good understanding of real catalytic behaviour; real catalyst particles contain many steps, edges, corners and missing atom defects and these can be modelled by studying stepped and kinked single crystal surfaces, such as those shown in Fig. 2.8. These surfaces are in fact high indexed faces of the single crystal.

2.2 Metallic bonding

In this section we will look at how bonding in metals can be described. Bonding is of great importance for determining many aspects of adsorption on metal surfaces. We can begin by considering the molecular orbitals which result from bringing two atoms

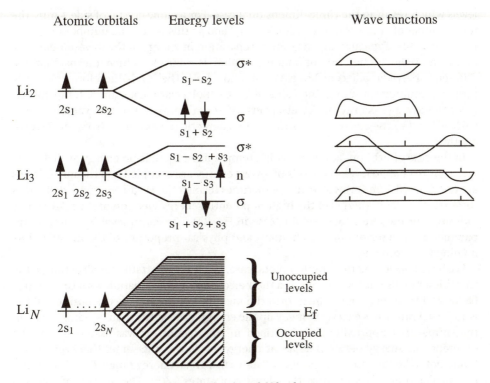

Atomic orbitals   Energy levels   Wave functions

N energy levels and *N* 2s electrons

Fig. 2.9 Metallic bonding in one dimension. This figure shows the molecular orbitals formed from 2, 3 and *N* atomic orbitals of Li, and how they are occupied by the available electrons. The Fermi level E_f is indicated for the array of *N* Li atoms. The wavefunctions for each of the resulting molecular orbitals are also shown.

together. In our example, lithium, which has electronic configuration $1s^2\ 2s^1$, produces σ-bonding and σ^*-antibonding orbitals as shown in Fig. 2.9, together with the associated wavefunctions. The number of electrons available to occupy these orbitals is two, as one is available from the $2s^1$ level of each of the two Li atoms. These electrons occupy the σ-bonding orbital; the σ^*-antibonding orbital remains empty. It is interesting to note that $Li_{2(g)}$ exists and has a bond energy of 106 kJ mol^{-1}.

We can now consider the case of Li_3, which is also shown in Fig. 2.9. In this case the three 2s orbitals combine to produce a σ-bonding orbital, a non-bonding orbital and a σ^*-antibonding orbital. There are now three electrons available and so two occupy the σ-bonding orbital, while the third occupies the non-bonding orbital.

We can build up a molecular orbital diagram for a one-dimensional metal by considering what happens if we now look at *N* Li atoms. For each atomic orbital, a new molecular orbital is created. Thus, for *N* atoms, *N* new levels are created, as shown in Fig. 2.9. There are *N* 2s electrons available to occupy these levels and they fill the lowest energy levels first, two electrons per orbital. Similar molecular orbital pictures can be deduced for a three-dimensional crystal. It is interesting to estimate the number of levels which result in the three-dimensional case. For a cubic crystal of side 1 mm, the total number of atoms is of the order 5×10^{19} and so this is also the number of energy levels available. Typically, the maximum separation in energy of the lowest σ-bonding orbital and the highest σ^*-antibonding orbital is $\sim$20 eV, which means that the difference between adjacent levels is $\sim 4 \times 10^{-19}$ eV. The levels therefore effectively form a continuum of states. The value of kT (the Boltzmann constant $\times$ temperature) gives the average energy of the electrons. At room temperature this value is about 0.026 eV and so the electrons are sufficiently energetic to pass easily from one level to another.

In the case of lithium, at 0 K, at which temperature there is no energy available for electrons to change level, the levels of lowest energy are occupied and the upper levels are unoccupied. The Fermi level (E_f) (sometimes termed the Fermi energy) is defined as the energy level (or energy) of the highest potential energy electron in the system. It is indicated for the one-dimensional Li case in Fig. 2.9. The Fermi level is an important parameter for determining the chemical and physical properties of the metal and its binding to adsorbates.

We have considered the simplest metal case, but it is important to realize that in the transition metals it is the d-bands and their electrons, in combination with the s- and p-bands and their electrons, which form the metallic continuum, so the position of the Fermi level rarely lies exactly halfway up the energy scale at the 'non-bonding' level. In the simple first approximation described in this section, it was assumed that the difference in energy between adjacent energy levels was the same. However, this is found not to be the case. The number of states in a given energy range, $E \rightarrow E + \mathrm{d}E$, can be calculated and is known as the 'density of states', $Z(E)$. The density of states is dependent on both the number of states at a given energy and the separation of states within the energy region over which the density is being determined. Calculations of the density of states show that at higher energies the energy levels are closer together and so the density of states gets larger at higher energies. The density of states is discussed in more detail in Section 2.2.3.

It should also be noted that any interaction with adsorbates can cause an interchange of electrons between the adsorbate and the metal, and so shift the Fermi level up or down.

2.2.1 The jellium model

Calculation of the electronic states in a solid is complicated because the most weakly bound electrons are delocalized, that is they can move freely around the ion cores. They therefore experience a periodic potential. A further problem for the theorist is that the free electrons repel each other. This means that to get a complete description of the electronic states, Schrödinger's equation has to be solved for a many-body periodic potential and this is rarely possible. Several different approximations are therefore made to allow comparison to be made with experiment.

One of the least complicated models is the free electron model, which approximates the ion cores in their periodic arrangement to a smeared-out, uniform, attractive potential. The smeared-out potential applies to a fictitious material which is called a jellium. The positive potential is modelled by cutting off the surface abruptly. The form

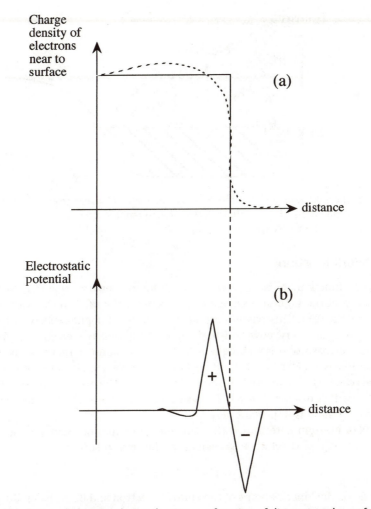

Fig. 2.10 (a) Variation of electron charge density as a function of distance to the surface (dashed line). (b) the associated electrostatic potential (the surface dipole moment) (solid line).

of the variation of the electron density in a free electron metal with a single outer electron per atom is shown in Fig. 2.10(a). The ion cores produce a positive potential at the surface. However, because the electrons are 'free' in the solid, they move towards the surface to screen out this positive potential. This produces both an increase in the electron density near to the surface and an oscillation in electron density (the origin of which is somewhat complex and concerns the interference of the wave properties of the electrons), which dies away rapidly into the solid. The wavefunctions of the electrons spill out into the vacuum level as shown in Fig. 2.10(a) and this leads to a surface dipole moment (shown in Fig. 2.10(b)), which contributes to the so-called 'work function' of the surface.

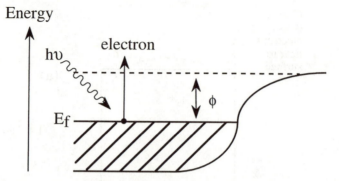

Fig. 2.11 The work function, ϕ, is the difference in energy of an electron at the Fermi level and that just outside the surface. A photoelectron being emitted from a surface must over come this energy in order to escape.

2.2.2 Work function

The work function, ϕ, is defined as the minimum energy required to remove an electron from the highest occupied energy level in the solid to the 'vacuum level' as shown in Fig. 2.11. This has important implications for photoelectron spectroscopy (PES), which is a very powerful tool for investigating the electronic properties of surfaces. The basis of this technique is that when a 'monoenergetic' X-ray or UV beam of radiation is incident on a solid surface, electrons which have particular kinetic energies can be photoemitted, as shown in Fig. 2.12. The origin of these photoemitted electrons will be discussed in detail in Section 2.2.4; however, the influence of the work function on them is important here.

In 1900, Einstein determined that the energy of the incident radiation $h\upsilon$ can be found from eqn (2.1), which is known as the Einstein relation:

$$h\upsilon = E_{KE} + BE \tag{2.1}$$

where E_{KE} is the kinetic energy of the emitted electron and BE is the binding energy of the electron within the solid. This means that by irradiating a surface with photons of sufficiently high energy, we can measure the kinetic energy of the resulting electrons and thus determine the binding energy of those electrons. However, the kinetic energy of the photoelectrons emitted from the surface is reduced by the work function. The Einstein relation in eqn (2.1) must be modified to give

$$h\upsilon = E_{KE} + E_B + \phi \tag{2.2}$$

So the binding energy of the electrons in the solid, BE, is the sum of the binding energy of a core level with respect to the Fermi level, E_B, plus the work function ϕ, as indicated in Fig. 2.12.

The work function is obviously different for different metals because each has a different range of energies in the band structure and different Fermi energies. However, a typical value for a metal is about 5 eV.

The work function can be written in terms of the Fermi energy, since this is the level of the electrons at the top of the valence band. The expression in eqn (2.3) relates ϕ to

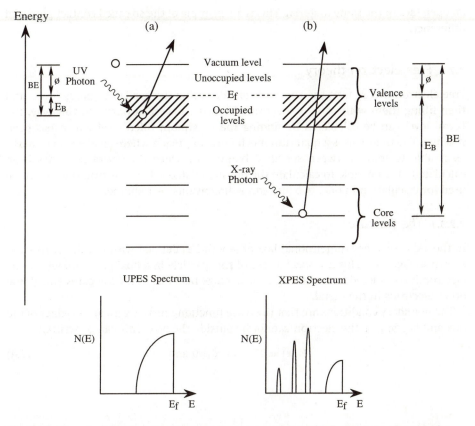

Fig. 2.12 The principle of photoelectron spectroscopy. (a) A UV photon is incident on the sample and ejects a valence electron. (b) An incident X-ray photon may eject a valence electron but more commonly ejects a core level electron as it has sufficient energy to do this. In both UPES and XPES the electron overcomes the binding energy BE in order to escape and is released with kinetic energy E_{KE}. The sum of these two components (BE $+ E_{KE}$) is equal to the incident photon energy. BE is equal to the sum of the binding energy with respect to the Fermi level plus the workfunction (E_B and ϕ). BE, E_{KE} and ϕ are indicated on the diagram for both (a) and (b). Sketches of the expected form of the spectra, plotted as $N(E)$ against E, are shown below the energy level diagrams.

the charge on the electron, e, the potential due to coupling with valence electrons in the bulk electron density, V_{ex}, and the surface 'space-charge' potential which an electron has to overcome to escape the surface, V_{dip}.

$$\phi = V_{ex} + V_{dip} - E_f/e \qquad (2.3)$$

The V_{dip} part of the expression shows why each different surface of a given metal has a different work function. The reason is that there are different electron densities at different surfaces, so close-packed faces (e.g. fcc (111)) generally have higher work functions than more open ones (such as (100)); points just outside the surface of a closed-packed face have a negative electrostatic potential relative to points just outside an open face of the same material. This is an example of the so-called contact potential difference.

2.2.3 Free electron theory

Free electron theory uses the approach of calculating the available energy levels and then filling them with pairs of electrons from the lowest upwards. In this way, the Fermi level can be calculated by finding the maximum filled level which has been reached. The theory uses a quantum mechanical approach where an electron of mass, m, is able to move between specified boundaries. Once the Fermi level has been calculated, it is possible to calculate the density of states. Free electron theory can be used for calculation in one, two and three dimensions, as required.

2.2.3.1 *The 1-d case*

In the 1-d case, a one-dimensional box of length L is defined with infinite barriers as shown in Fig. 2.13. This is a special case of the 'particle in a box' problem, where the electron is considered to be moving in an average field due to the ion cores and all the other electrons in the metal.

The boundary conditions are that the wave function must be zero at the edges of the box and beyond, as the electron cannot be outside the box. This can be written

$$\psi_n(x) = 0 \qquad \text{at } x = 0 \text{ and } x = L \tag{2.4}$$

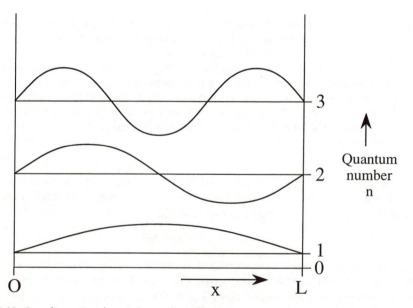

Fig. 2.13 One dimensional particle-in-a-box showing the wavefunctions applicable to the energy states available for electrons in a 1-d metal. In this case the box has infinite barriers.

Potential energy can be neglected in this case and so we can consider the electron's kinetic energy alone. Thus the Schrödinger equation reduces to:

$$H\psi = \left(\frac{p^2}{2m}\right)\psi \tag{2.5}$$

where H is the Hamiltonian and p is the momentum operator, $-\frac{\hbar}{i}\frac{d}{dx}$. E_n, the energy of an electron in an orbital of quantum number n, can then be calculated from

$$H\psi_n(x) = -\frac{\hbar^2}{2m}\frac{d^2\psi_n(x)}{dx^2} = E_n\psi_n(x) \tag{2.6}$$

The boundary conditions are satisfied if we take a solution of the form

$$\psi_n(x) = A\sin\left(\frac{2\pi x}{\lambda_n}\right) \quad \text{where} \quad \frac{1}{2}n\lambda_n = L \tag{2.7}$$

A is a constant and all λ_n are the wavelengths of the allowed solutions. Thus

$$\psi_n(x) = A\sin\left(\frac{\pi n x}{L}\right) \tag{2.8}$$

Because

$$\frac{d^2\psi_n(x)}{dx^2} = -A\left(\frac{n\pi}{L}\right)^2\sin\left(\frac{\pi n x}{L}\right) \tag{2.9}$$

eqns (2.6) and (2.9) can be combined to give

$$E_n = \frac{\hbar^2}{2m}\left(\frac{n\pi}{L}\right)^2 \tag{2.10}$$

These levels are filled up from the lowest energy first, by inserting two electrons of opposite spins (remembering the Pauli exclusion principle) into each level.

2.2.3.2 *The Fermi level in 1-d*

The strict definition of the Fermi level (E_f) is the energy of the highest occupied level at 0 K. If there are N_f valence electrons, then the quantum number for this level must be

$$n_f = \frac{N_f}{2} \tag{2.11}$$

where n_f is the quantum number of the Fermi level. From eqn (2.10) the energy of this level E_f is

$$E_f = \frac{\hbar^2}{2m}\left(\frac{n_f\pi}{L}\right)^2 = \frac{\hbar^2}{2m}\left(\frac{N_f\pi}{2L}\right)^2 \tag{2.12}$$

The Fermi level is calculated here with respect to the bottom of the band of energy levels.

2.2.3.3 *The 3-d case*

The treatment described above can be extended to encompass three dimensions (*x,y,z*). In this case the electron is now confined in a three-dimensional cube of edge *L*. Again, the electron can move in the field which results from the average of the other electrons plus the ion cores. As the electron's position is confined to within the box, ψ is zero at all edges of the cube and so the solution to the Schrödinger wave equation is analogous to that in the one-dimensional case (eqn (2.8)).

$$\psi_{n(r)} = A \, \sin\left(\frac{n_x \pi x}{L}\right) \, \sin\left(\frac{n_y \pi y}{L}\right) \, \sin\left(\frac{n_z \pi z}{L}\right) \tag{2.13}$$

where A is a normalizing constant. From eqn (2.10), equating the expression for E_n in each direction, the total energy E can be calculated.

$$E = E_x + E_y + E_z = \frac{\hbar^2 \pi^2}{2mL^2} \, (n_x^2 + n_y^2 + n_z^2) \tag{2.14}$$

or

$$E = \frac{\hbar^2 n^2}{2mL^2} \tag{2.15}$$

where n^2 is the sum of the squares of the n values associated with the orthogonal directions (x, y, z).

When working in three dimensions it is often convenient to work in momentum space (sometimes called k-space), where $\underline{k}$ is a wavevector of an electron and the momentum $p = \hbar \underline{k}$. $\underline{k}$ is in the direction of the wave (which is perpendicular to the wavefront) and its magnitude is given in relation to the wavelength, λ, by

$$|\underline{k}| = 2\pi/\lambda \tag{2.16}$$

$\underline{k}$ can be resolved into three components k_x, k_y and k_z. Analogy with the 1-d case (eqn (2.7)) gives

$$\lambda_n = \frac{2L}{n} \quad \text{and} \quad k_x = \frac{n_x \pi}{L_x}$$
$$k_y = \frac{n_y \pi}{L_y} \tag{2.17}$$
$$k_z = \frac{n_z \pi}{L_z}$$

Combining eqns (2.13) and (2.17) gives

$$\psi_n = A \, \sin(k_x x) \, \sin(k_y y) \, \sin(k_z z) \tag{2.18}$$

Substitution of eqn (2.18) into the three-dimensional form of the Schrödinger wave equation leads to

$$E_{\underline{k}} \psi_{\underline{k}} = -\frac{\hbar^2}{2m} \left(\frac{\partial^2}{\partial x^2} + \frac{\partial^2}{\partial y^2} + \frac{\partial^2}{\partial z^2}\right) \psi_{\underline{k}} \tag{2.19}$$

where $\underline{k}$ denotes the set of numbers k_x, k_y, k_z. E depends only on the magnitude of $\underline{k}$. The value of E_k is then given by

$$E_k = \frac{\hbar^2 k^2}{2m} = \frac{\hbar^2 (\underline{k} \cdot \underline{k})}{2m} = \frac{p^2}{2m} \tag{2.20}$$

where $k^2 = k_x^2 + k_y^2 + k_z^2$.

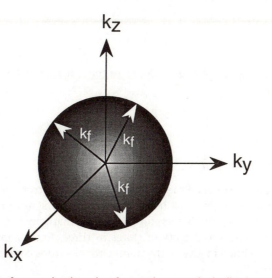

Fig. 2.14 The Fermi surface can be thought of as a sphere at which all values of k are the same. The sphere at radius k_f, the Fermi surface, separates the occupied and unoccupied states.

2.2.3.4 *The Fermi energy in three dimensions*

In the previous sections, the standing wave approach with boundary conditions driving the wave function to be zero at the edges of the box has been used to derive expressions for the energy states and the Fermi energy in one dimension. This approach can also be used to find the Fermi energy in three dimensions. However, a running wave picture with periodic boundary conditions can also be used and because either approach may be encountered in the literature, the running wave approach will be given here to derive an expression for the Fermi energy in three dimensions. The results from the two methods are identical. (It should also be noted that in addition to giving a more satisfactory picture of the periodicity of the lattice, the running wave approach is mathematically simpler to use.)

In the running wave scheme both the positive and negative values of k_x, k_y and k_z are allowed, corresponding to waves travelling in different directions, but the spacing between allowed k_x, k_y and k_z values is now $2\pi/L$ (cf. eqn (2.17)). This means that in **k**-space there is one allowed quantum state in a volume element of $(2\pi/L)^3$. One can imagine a spherical surface in **k**-space in which all points are at the same energy, as shown in Fig. 2.14. In the figure, the sphere corresponds to the energy E_f. The *Fermi surface* is that special sphere at radius k_f which separates occupied and unoccupied states at 0 K. The occupied volume in **k**-space is then $4\pi k_f^3/3$.

Thus N_f, the total number of electrons within the Fermi surface, can be found from

$$N_f = 2\frac{4\pi k_f^3/3}{(2\pi/L)^3} = \frac{L^3}{3\pi^2}\,k_f^3 \tag{2.21}$$

The factor 2 takes into account the fact that the states are each occupied by one spin-up and one spin-down electron. For volume $V' = L^3$,

$$k_f = \left(\frac{3N_f \pi^2}{V'}\right)^{\frac{1}{3}} \tag{2.22}$$

Substitution of eqn (2.22) into eqn (2.20) gives the value for E_f, the energy of the Fermi level:

$$E_f = \frac{\hbar^2}{2m} \left(\frac{3N_f \pi^2}{V'}\right)^{\frac{2}{3}} \tag{2.23}$$

2.2.3.5 *The density of states*

The density of states, $Z(E)$, is defined as the number of electrons per unit energy range, and $Z(E)\,dE$ can be thought of as being the number of states present in the energy region from $E \to E + dE$; $Z(E)$ is therefore described mathematically as dN/dE.

In order to calculate the density of states in three dimensions, eqn (2.23) can be rearranged and modified to give N, the number of electrons contained within a sphere of energy E, in the form

$$N = \frac{V'}{3\pi^2} \left(\frac{2m}{\hbar^2}\right)^{\frac{3}{2}} E^{\frac{3}{2}} \tag{2.24}$$

Equation (2.24) can be used to determine N for any value of E (whereas, strictly speaking, eqn (2.23) applied specifically to the number of electrons in the Fermi surface). Differentiation of eqn (2.24) gives

$$Z(E) = \frac{dN}{dE} = \frac{V'}{2\pi^2} \left(\frac{2m}{\hbar^2}\right)^{\frac{3}{2}} E^{\frac{1}{2}} \tag{2.25}$$

A plot of $Z(E)$ against E is a parabolic function which is shown in Fig. 2.15. At a temperature of 0 K all levels below the Fermi level are filled and those above E_f are unoccupied, as shown in the figure. As the temperature is increased so that $T > 0$ K, thermal excitation allows a tiny fraction of the electrons in the highest levels, those within a few kT of E_f, to spill upwards into the unoccupied levels above E_f.

2.2.4 Experimental evidence for the density of states—photoelectron spectroscopy (PES)

The most compelling evidence for the existence of the band structure and the density of states comes from PES. The principle of PES is shown in Fig. 2.12.

Figure 2.12(a) shows an electron being ejected from the valence band below the Fermi level. Ejection of an electron from the valence band results from excitation by UV photons (UPES), typically in the 5–20 eV energy range, while core electrons are excited by X-rays (XPES), typically in the 10–1000 eV energy range as shown in Fig. 2.12(b). X-ray photon excitation may also result in ejection of electrons from the valence band. The form of the spectrum obtained when using UV and X-rays as excitation sources is shown schematically in Fig. 2.12. The photoelectrons ejected from the core levels carry valuable fingerprint information on the solid/surface under investigation.

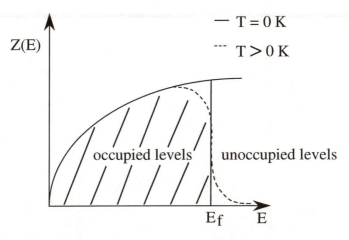

Fig. 2.15 A plot of $Z(E)$ against energy, E, reveals that the density of states is a parabolic function. The Fermi level at 0 K is indicated by a solid line and the effect on the occupation of states of raising the temperature above 0 K is shown by a dashed line.

Figure 2.16 shows the XPES spectrum of a single crystal of copper. The major peaks can all be correlated with the ionization energies of the atomic orbitals of Cu, as indicated on the figure. The sloping background on which the peaks appear arises from the emission of secondary electrons from collisions of the emitted electrons with the

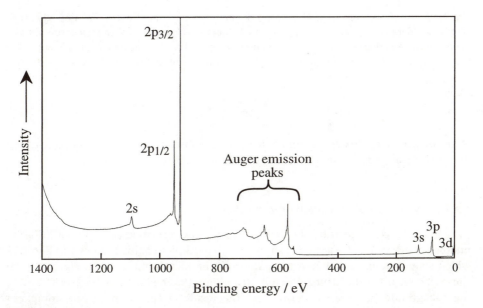

Fig. 2.16 An XPES spectrum of Cu. The orbitals from which the photoelectrons arise are indicated on the figure. (Redrawn from the *Handbook of X-ray photoelectron spectroscopy*, Perkin-Elmer Corporation, 1992).

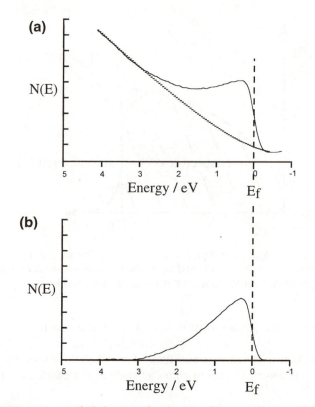

Fig. 2.17 An UPES spectrum of Al showing the density of occupied states d$N(E)$/dE against E: (a) the raw data where the curve appears on a sloping background; (b) the background has been subtracted and the resulting curve can be compared to the calculated curve in Fig. 2.15. (Spectrum recorded by Dr S.A. Morton of the Department of Chemistry, University of York.)

atoms (see Section 2.2.4.3). Additional features are observed which are due to so-called Auger electron emission and their origin will be discussed in Chapter 3.

2.2.4.1 Density of states measured by PES

UPES can be used to measure the density of states by recording the value of $N(E)$, the number of electrons of energy E, against E. A typical plot of this is shown in Fig. 2.17 which shows the occupied density of states for an aluminium sample. Figure 2.17(a) shows the recorded spectrum, which is on a highly sloping background. In Fig. 2.17(b) the background has been removed and the resulting spectrum can be compared directly with the theoretically predicted plot of $Z(E)$ against E which is shown in Fig. 2.15. The similar shape and form of the plot in Fig. 2.17(b) gives direct evidence of the existence of the density of states.

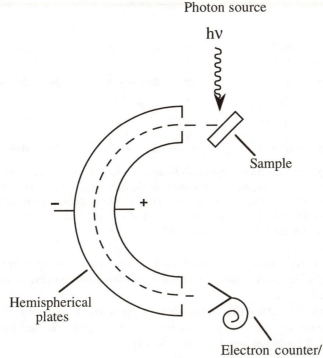

Photon source

hν

Sample

− +

Hemispherical
plates

Electron counter/
multichannel detector

Fig. 2.18 A schematic diagram of a concentric hemispherical analyser (CHA)—typical apparatus for recording PES spectra. Photons are incident on the sample and the photoelectrons emitted enter the CHA. The two hemispheres guide the electrons towards the detector. They are set at a constant difference in potential with the outer hemisphere set negative with respect to the inner hemisphere. A potential is applied to both hemispheres and this is swept across the desired energy range to detect photoelectrons in each resolution element. The electrons are counted by a multichannel detector or an electron counter.

2.2.4.2 *The PES experiment*

The technology required for electron energy analysis was developed in the 1960s. Typical apparatus for carrying out PES measurements is shown in Fig. 2.18.

(i) The source
For UPES measurements a high-intensity UV lamp can be employed. For XPES a typical X-ray source is produced by bombarding a metal target with electrons; one of the most common target materials is aluminium. Figure 2.19 shows the effect of a 30-keV electron on the aluminium target. The incident electron has sufficient energy to knock out one of the 1s core-level electrons in (a), leaving a very unstable Al^+ ion, as shown in (b). The Al^+ ion may relax when a higher level electron (in this case the 2p electron) drops down to the 1s level, emitting energy $h\nu$, which is the difference between the binding energies of the two levels. This process is called X-ray fluorescence (XRF) and

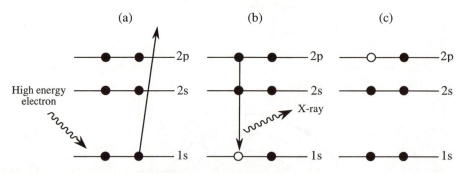

Fig. 2.19 The principle of X-ray fluorescence. (a) An incident high-energy electron ejects a core-level electron from the metal atom (the 1s level in the case of Al shown here). (b) The resulting unstable species is partially stabilised when a higher-lying electron drops back to the 1s level (in the Al case this is from the 2p level). The energy evolved is released as an X-ray photon. (c) shows the final state formed which is Al^+.

has found wide application in the analysis of solids. In the case of the 2p–1s transition of Al^+, the value of $h\nu$ is given a special symbol $k_\alpha = 1486.6$ eV and this is used as a standard for solids.

The source is separated from the ultrahigh vacuum (uhv) chamber by a thin Al window, which is necessary because the vacuum in the X-ray source is quite poor. The X-rays pass through the window and then bombard the surface.

(ii) The analyser
Electrons emitted from the surface then pass into an energy analyser. There are four methods of electron energy analysis following photoemission. These use the magnetic double focusing spectrometer, the electrostatic retarding potential spectrometer, the electrostatic cylindrical mirror analyser (CMA) and the electrostatic concentric hemispherical analyser (CHA). Figure 2.18 illustrates the CHA which is the one described here.

The electrons ejected from the sample enter the hemispherical analyser where a constant potential difference is maintained between the inner and outer hemispheres. The inner hemisphere is kept at a more positive potential than the outer, as the electrons are being guided along a curved path. A retarding potential relative to the sample is applied to both hemispheres to allow only electrons above a desired energy to enter the analyser. The retarding potential is swept across the range of energies of interest to record the spectrum. The resolution is dependent on the difference in potential across the two hemispheres.

(iii) The detector
Detection is usually carried out using an electron counter such as a channel electron multiplier or a channel plate. Recent developments in electron energy analysis and detection have produced multichannel spectrometers in which electrons of several different energies are detected simultaneously. These spectrometers rely on the use of multichannel electron detectors; the advantage of using multichannel spectrometers is that spectra can be recorded very rapidly over the desired energy range; the disadvantage is that they are expensive and complex.

2.2.4.3 *The inelastic mean free path*

When discussing PES or any other type of surface spectroscopy where electrons penetrate or are ejected from a surface, it is important to bear in mind the concept of the *inelastic mean free path*, λ, which can be described as the average distance travelled by an electron through a solid without losing energy. A plot of λ against electron energy is reproduced in Fig. 2.20; this is sometimes called the 'universal curve'. The plateau region of the curve at around 50 eV coincides with electron energies typical of PES experiments.

Energy loss by electrons can occur in a number of ways but principally by

(i) Exciting electrons in other atoms from the ground state to an empty state in the atom;
(ii) Ionizing another atom, causing ejection of a further photoelectron (for a lower energy transition);
(iii) Interaction with, or excitation of, a bulk or surface plasmon. A plasmon is a collective density fluctuation of the electron gas in the solid. The plasmon is quantized with energy in the range 5–25 eV, depending on the density of the electron gas. A bulk plasmon is in three dimensions whereas a surface plasmon is in two dimensions.
(iv) Excitation of a lattice vibration or phonon mode. Energy losses of this type are very small—a few tens of meV.

The shape of the universal curve is governed by the energy loss processes that are available for a given electronic energy. For electron energies below 30 eV, λ increases rapidly, because electrons of less than 30 eV are below the critical energy for plasmon excitation. Above electron energies of 50 eV, it is found that λ is proportional to the

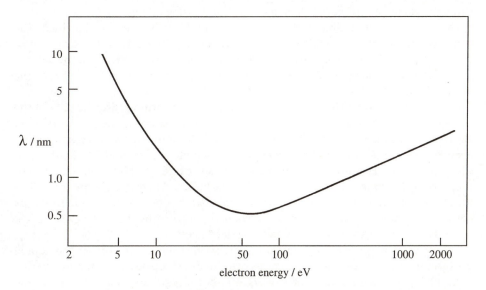

Fig. 2.20 The 'universal curve' showing the inelastic mean free path (λ) for electrons with a range of energies, in different metals. Note the logarithmic scales. (From Somorjai 1981.)

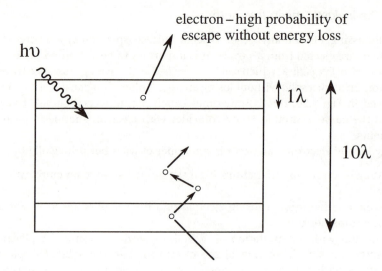

Fig. 2.21 Schematic diagram showing the likelihood of electrons escaping from a depth of 1λ and 10λ below the surface.

electron velocity, because as the electron travels faster, it takes less time to pass through a solid and so there is less likelihood of it losing energy.

The 'universal' applicability of the curve to most solids is easily explained. The dominant mechanism by which electrons lose energy in solids is via excitation of electrons in the valence band. For most materials the electron density in the valence band is constant at about 0.25 electrons per cubic angstrom, thus the energy loss to the valence band is similar in most materials.

Figure 2.21 illustrates the observation that an electron within a distance of 1λ of the surface is much more likely to escape from the surface without energy loss than an electron at a distance 10λ from the surface. It should be noted that X-rays can penetrate deep into the surface, so photoelectrons can be generated at considerable depth. If, on the other hand, excitation of the photoelectrons is caused by high-energy electrons, they too will be subject to interactions within the bulk and may lose energy and will be unable to penetrate the sample to considerable depth. The depth of penetration by both incident and emitted electrons is obviously governed by their energy E, since λ depends on E as shown in Fig. 2.20 and described above.

The effect of energy losses by the processes described above is to produce the sloping background on the spectra noted earlier, the so-called *secondary electron background*. Thus PES peaks appear as sharp peaks on the background slope; however, the background is often removed either electronically or by computational means in reported spectra.

2.3 Surface energies

In Chapter 1, the reason for the reactivity of surfaces was described in terms of the higher energy at the surface in comparison to that of the bulk sample. The surface energy can be estimated very crudely by using a measurement of the latent heat of vaporization of the solid, ΔH_v. In the bulk, a metal atom has 12 nearest neighbours in an fcc crystal, so when an atom evaporates from the surface, six bonds are broken on average. (This takes an average over all the different possible faces, for example removing an atom from a threefold site on top of the (111) face requires that three bonds are broken while removing one from within the top (111) layer requires the breaking of nine bonds, etc.)

If we make the approximation that only nearest neighbour interactions are important and that the interactions with next nearest neighbours are very small, we can estimate the value of the energy E_b required to break each M–M bond as

$$E_b = \frac{\Delta H_v}{6L} \tag{2.26}$$

where L is the Avogadro number. If we were to separate the bulk and form two (111) planes each with N_s surface atoms, then three bonds would be broken per surface atom, considering that this is like removing a whole plane of isolated atoms from on top of the (111) plane. The surface energy, γ, is therefore given by

$$\gamma = \frac{3\,E_b\,N_s}{2} \tag{2.27}$$

where the factor 3 accounts for the breaking of three bonds per surface atom, and the factor 2 accounts for the formation of two (111) surfaces. Substituting eqn (2.26) into eqn (2.27) gives a value for γ in terms of the heat of vaporization.

$$\gamma_{(111)} = \frac{3}{2 \times 6}\,\frac{\Delta H_v}{L}\,\frac{4}{\sqrt{3}a^2} = \frac{\Delta H_v}{L\sqrt{3}a^2} \tag{2.28}$$

where $N_s = 4/\sqrt{3}a^2$ for the (111) surface, as shown in Section 2.1.3. Obviously, different unit cell dimensions and numbers of bonds being broken must be taken into account when calculating γ for other faces. In the (100) case, four bonds are broken per surface atom and $N_s = 2/a^2$ (again as shown in Section 2.1.3). Thus

$$\gamma_{(100)} = \frac{4}{12}\,\frac{\Delta H_v}{L}\,\frac{2}{a^2} = \frac{3\Delta H_v}{4La^2} \tag{2.29}$$

for a given crystal, the relative surface energy can be estimated from the expressions for each specific surface and are found to be in the order

$$\gamma(111) < \gamma(100) < \gamma(110) < \gamma \text{ (stepped faces)}$$

showing that more compact faces have lower energy and are therefore less reactive.

2.4 Bonding in semiconductors

2.4.1 Breakdown of the free electron theory

Although the free electron theory is excellent for modelling the behaviour of many metals, it is found to break down if applied to all elements as we move across the periodic table. This is because the prediction for electron density (N/V) is a steady increase across the periodic table, implying an increase in the density of states at the Fermi level. Of course, this would tend to imply that an element such as carbon should be an excellent conductor in any of its three forms. However, as it is an electrical insulator when tetrahedrally bonded in the diamond structure, we have a perfect example of the breakdown of the theory. Obviously, this means that bonding in semiconducting, semi-metallic and insulating materials must be considered using a different approach.

In this chapter the discussion is mainly restricted to metals and semiconductors, which are of particular interest to surface scientists. Discussion of semiconductors will only encompass those features which are of importance to the properties of the surface.

2.4.2 Covalent bonding

In bulk semiconductors, the structures are largely the result of covalent bond formation. In particular, where hybrid orbitals are formed from the intermixing of s and p orbitals to form sp^3 orbitals, particularly strong bonding is obtained. These sp^3 orbitals are highly directional and result in the two most commonly found semiconductor crystal structures, namely the diamond and zinc blende (α-ZnS) structures which are shown in Fig. 2.22(a) and (b), respectively.

A consequence of the free electron theory is that there is no upper limit to the energy of the band of states because $E \propto |\mathbf{k}|^2$, which certainly does not correspond well with our original predictions using molecular orbital theory (see Section 2.2). If we plot energies of allowed states against $\mathbf{k}$ (when the periodic potential of the lattice is included), we find that the band of energies is, in fact, finite as shown in Fig. 2.23. This example shows the actual states for a 1-d array of H atoms (the solid curve) and the states predicted using free electron theory (dashed line). At low $\mathbf{k}$ the real behaviour is in good agreement with that of the free electron model, but at higher values, clearly, it is not. For other more complex systems involving p and d orbitals, it is often found that the variation of E with $\mathbf{k}$ (known as the dispersion) is vastly different from the s-band dispersion curve of Fig. 2.23. For example, it is possible for the dispersion curve to have a maximum at $\mathbf{k} = 0$. Because each orbital generates an associated band, the dispersion of the bands is controlled by the orbital overlap in each specific case. The greater the overlap, the greater the band width observed.

For Na, the correlation between the charge distribution (in real space) and the dispersion curves (reciprocal space) can be easily seen. For Na, which has the structure $1s^2 \, 2s^2 \, 2p^6 \, 3s^1$, there is significant overlap of the charge distributions for the 3s electrons. The 1s, 2s and 2p electrons remain localized at each Na nucleus and are not

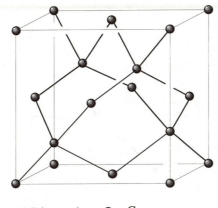

(a) Diamond   ● - C

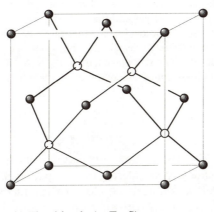

(b) Zinc blende (α-Zn-S)

● - Zn   ○ - S

Fig. 2.22 The two most common semiconductor crystal structures: (a) diamond and (b) zinc blende (α-ZnS).

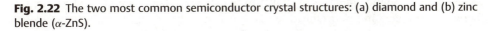

expected to interact significantly. As a result the dispersion curves for 1s, 2s and 2p electrons, the 'core electrons', remain as tightly defined and narrow energy levels. The overlap of the 3s electron charge distributions leads to the expectation that the 3s electrons will 'delocalize' from the atomic cores, thereby losing the discrete character of their energy levels. As discussed for the case of Li, we therefore expect the discrete 3s levels to blend to form a band structure. The resulting dispersion for the 3s level is shown schematically in Fig. 2.24. The Fermi level is indicated and shows that the 3s band is half filled and that, therefore, Na is a good conductor. Good conductivity occurs when only a slight increase in energy is required for the conduction electrons to move into the available vacant energy levels.

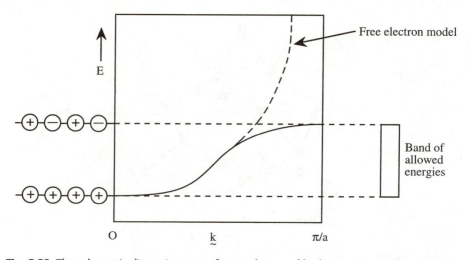

Fig. 2.23 The schematic dispersion curve for a 1-d array of hydrogen atoms. The solid curve shows the actual band of allowed energies and the dashed line shows the states predicted on the basis of the free electron model. At low **_k_**, agreement between the two is good. At high **_k_**, free electron theory predicts an infinite array of states but in fact a band of allowed energies is observed. The orbitals involved are illustrated schematically to the left of the figure; the bandwidth is dependant on the extent of overlap between them. (The greater the overlap, the greater the bandwidth).

2.4.3 Conductivity

The amount of conductivity displayed by a solid falls into one of three main categories, and the behaviour is explained by considering the occupation of energy levels as shown in Fig. 2.25. For the conductors shown in (a), electrical conductivity is achieved readily because the bands are partially filled with electrons. This can either arise simply (as in the case of Na), or from the overlap of bands, for example, the s and p orbitals as shown schematically in Fig. 2.26.

The case for semiconductors is shown in Fig. 2.25(b). Here a small band gap exists between occupied and unoccupied levels which may be induced by the presence of impurities, such as that illustrated in (b)(ii). The small band gap enables electrons with sufficient thermal energy (kT) to cross the gap easily (see Section 2.2).

The final part of Fig. 2.25(c) shows the requirement for an insulator. A large band gap is required to inhibit the flow of electrons from the occupied to the unoccupied levels, so that they have insufficient energy to cross the gap. It is possible to excite electrons to cross these large energy barriers using excitation by highly energetic radiation such as UV photons or X-rays.

The point at which a solid is defined as a conductor, or a semiconductor, is determined largely by the overlap of adjacent levels. For example, Mg conducts, although consideration of the electronic structure, $1s^2\ 2s^2\ 2p^6\ 3s^2$, implies that it should be insulating (because all the orbitals are filled). However, there is significant overlap of the filled 3s orbitals, with the unoccupied 3p orbitals which become partially filled as shown in Fig. 2.26. Thus Mg is a good electrical conductor.

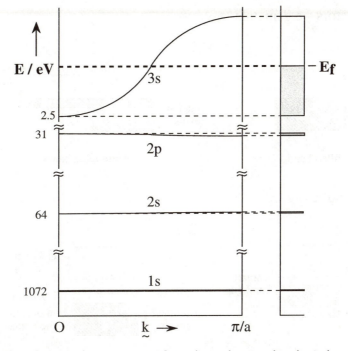

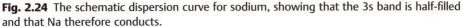

Fig. 2.24 The schematic dispersion curve for sodium, showing that the 3s band is half-filled and that Na therefore conducts.

For the first-row transition metals it is important to note that the d-band (nd) overlaps the s-band in the next level (($n + 1$)s), and both are partially filled, which is a key reason for the vast range of reactivity demonstrated by transition metal surfaces. In noble metals such as copper, the d-band is completely filled and the s-band is partially filled, giving rather different surface reactivity.

2.4.4 Electronic structure of semiconductors

The theory most commonly used to describe bonding in a semiconductor lattice takes a similar approach to that of the free electron theory, but also takes into account the directionality of bonding within the lattice by considering the symmetry conditions when obtaining the linear combinations of the wavefunctions which describe the orbitals.

A detailed treatment of this topic will not be given here, as it is more appropriately discussed in textbooks describing the solid state. However, it should be noted that the electronic structure is largely influenced by orbital overlap. In addition, for a surface where an unsatisfied valency exists, all of the energetic considerations applied to metal surfaces should also be applied to semiconductors. In other words, the unsatisfied valency leads to a relatively high surface energy.

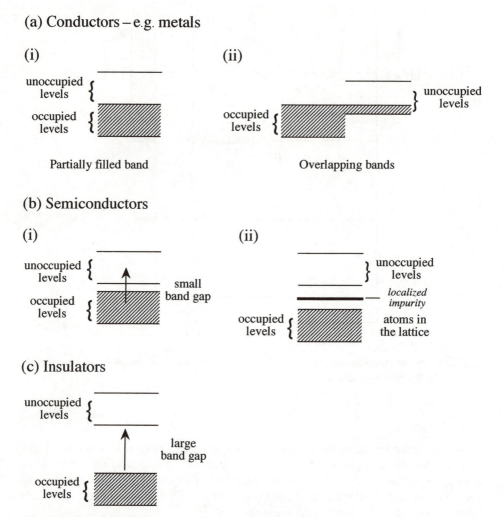

Fig. 2.25 The origin of electrical conductivity. (a) shows the energy levels required for conduction: (i) for a partially filled band and (ii) for overlapping bands. (b) (i) illustrates the requirement of a small band gap (of the order 0.1–2 eV at room temperature) for semiconducting materials; (ii) the band gap may be induced by the presence of impurities. (c) The large band gap required for an insulator (of the order several eV at room temperature).

2.4.5 Bonding

From the foregoing discussion, it is obvious that at 0 K, a semiconducting crystal will not conduct electricity. It becomes semiconducting at higher temperatures because of thermal excitations over the small band gaps that are introduced by impurities, defects or mixed valence. In general, there are two types of semiconductors.

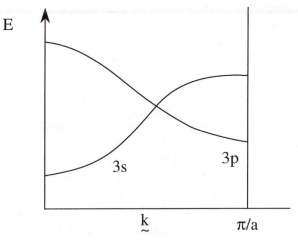

Fig. 2.26 The possible overlap of s and p orbitals.

2.4.5.1 *Intrinsic semiconductors*

Intrinsic semiconductors are highly pure materials which only semiconduct at temperatures sufficient to promote electrons thermally from the occupied valence band to the unoccupied conduction band (i.e. where $kT \approx$ the band gap). In addition to the electrons which are promoted to the conduction band, electrons in the valence band and the mobile 'holes' they create in it contribute to the electrical conductivity of the semiconductor.

Considering the diamond structure of the binary Ge–Si system, the tetrahedral bonding is associated with sp^3 orbitals, formed from the 3s and 3p orbitals in Si and the 4s and 4p orbitals in Ge. In the solid these orbitals become bands and so the crystal can be likened to a covalently bonded molecule. These four bands split, forming two sets of states comprising a high-energy antibonding state and a lower-energy bonding state, with an energy or band gap between the two sets of states. The bonding orbitals which point between atoms are occupied, and the antibonding states which point away from nearest neighbours are unoccupied. The value of kT at room temperature is about 0.026 eV, so for an insulator the band gap is several eV; diamond, for example, has a band gap of 5.4 eV. Typical values for semiconductors are 0.7 eV for Ge, and 0.2 eV for InSb.

2.4.5.2 *Extrinsic semiconductors*

In these semiconductors, conduction is associated with the presence of impurities in the material. Impurities can be introduced by 'doping' the crystal to form either an n-type or a p-type semiconductor.

(i) n-type semiconductors
These are produced by doping the semiconductor with atoms which donate additional electrons into the conduction band. This can be done by doping a semiconductor from

Group IV with atoms from Group V in the periodic table, e.g. doping Ge with As. The impurity atom has five electrons available for bonding but only four are needed, so the remaining electron occupies a discrete energy level about 0.1 eV below the conduction band, where little thermal energy is required to promote it into the conduction band (see Fig. 2.25(b)).

(ii) p-type semiconductors

These are produced by doping the semiconductor with an atom which is electron deficient in comparison to the host material. Doping a semiconductor from Group IV with atoms from Group III in the periodic table, such as Ge doped with Ga, forms this type of semiconductor, where the charge is passed through the lattice via positive 'holes' that accept electrons. The energy of the acceptor falls between the conduction and valence levels as shown in Fig. 2.25(b)(ii). The energy gap between the top of the valence band and the acceptor level is about 0.01–0.1 eV.

The conductivity of an extrinsic semiconductor can be controlled precisely by doping levels.

2.5 Experimental considerations

2.5.1 Ultrahigh vacuum

Studies of single crystal surfaces are challenging, in part because of the need to study them under ultrahigh vacuum conditions. Kinetic theory can be used to show that at a pressure of about 10^{-6} mbar, a surface will be bombarded at a rate of about 10^{14}–10^{15} molecules cm^{-2} s^{-1}. From the calculations of surface density in Section 2.1.3, it is apparent that if the incident gas were reactive, the surface would become covered with gas molecules in only 1 second. In order to carry out any meaningful measurements on the surface, it is therefore necessary to study it under a vacuum in the range 10^{-10}–10^{-11} mbar, so that it can remain 'clean' for a few hours. Pressures in this region are described as being at ultrahigh vacuum (uhv) and the technology to attain and maintain these levels took a long time to develop.

Nowadays it is common to use apparatus in the form of large steel chambers, with equipment which is bolted on using 'knife edge' flanges that seal by biting into copper gaskets. The chambers are pumped out using a combination of vacuum pumps, such as oil diffusion pumps, turbomolecular pumps and cryopumps, backed by coarser mechanical rotary pumps. In order to attain uhv it is usual to bake the entire vacuum chamber for more than 48 hours, above 120°C, to remove contaminants, especially water, which tends to stick to the chamber walls and slowly degas into the chamber.

Both the vacuum level required and the need for bake-out imposes serious limitations on the materials which can be used in uhv. For example, special uhv compatible oils are required in the diffusion pumps because the vapour pressures of oils which are commonly used in diffusion pumps are much higher than 10^{-10} mbar. Some metals, such a Pb, have an inherently high vapour pressure and so cannot be used for making any components for use in uhv. The need for baking the chamber also precludes the use of certain materials, such as indium (which softens and may melt) and rubber seals for flanges. Where a flexible seal is required, such as in a rotatable flange or to seal against a soft window material such as KBr, a special form of rubber, known as Viton, is adopted. Viton must be baked well below 150°C, because above that temperature the plasticizer within it begins to sublime and the Viton gasket begins to lose its flexibility.

2.5.2 Sample cleaning

There are several possible ways in which to clean a sample in uhv and the method adopted depends very specifically on what the sample is. For example, tungsten can be cleaned by heating under specific, carefully controlled temperature and pressure conditions in oxygen. In this case, the conditions are such that WO_3 is not formed; the function of the oxygen is to react with impurities as they are brought up to the surface, and any oxide formed is removed by heating to a slightly higher temperature once the oxygen has been removed from the chamber. Other reactive gases can also be used in this way for preparing certain surfaces, for example hydrogen is often used to clean iron surfaces.

Other surfaces, such as silicon, can be cleaned simply by heating to high temperature in vacuum, following a suitable etch prior to mounting in uhv.

The most common cleaning method is by ion bombardment. In this method, the sample is bombarded with a beam of fairly energetic, inert gas ions, usually argon or neon, which 'sputter' off the top layers, removing both metal and contaminant atoms. The surface is then annealed to heal the craters produced by the sputtering process, and thus restore the order on the surface. Several cycles of the sputter–anneal process are required because a second effect of the anneal is to bring bulk contaminants up to the surface and these must be removed by further sputtering. (Other ions such as oxygen and gallium are sometimes used for sputtering but for other purposes which will be discussed in Chapter 3.) Single crystals of different materials often require many cycles of sputtering and annealing, depending on the bulk levels of impurities and the ease of segregation.

The 'usual' range over which surface science experiments are performed is from just above 77 K to the annealing temperature of the metal. This range is accessible because most experimental systems can use liquid nitrogen to cool the sample (mounted on some type of manipulator) cryogenically. At 77 K liquid nitrogen boils and the lowest temperature obtained in practice is usually a little higher than this. The slightly higher temperature results largely from the practical difficulties which are encountered when trying to provide good thermal contact between the sample and cooling system, while maintaining the ability to heat it to its annealing temperature at a reasonable rate. Liquid nitrogen is the cryogen of choice because in addition to making experiments accessible down to relatively low temperatures, it is widely available, readily stored and relatively cheap. It is possible, using specially designed manipulators, to use liquid helium as the cryogenic cooler. When these systems are used, the base temperature obtainable is usually somewhat higher than the boiling point of the helium, as it is again limited by heat gains from electrical and mechanical connections to outside the vacuum system, and their insulation.

2.6 Surface structure

Studies of the structure of surfaces have shown that most 'relax' in order to compensate for the unsatisfied valency produced on creation of the surface. The spacing between the surface layer and that just below it is reduced (usually significantly) from that found in the bulk. Generally, the lower the density and atomic packing, the smaller the distance between these adjacent layers. A comparison of close-packed and open surfaces of the same crystal shows that in the more open faces, the interatomic distances between outermost and second layer are smaller than in the close-packed surface. The perturbation caused by the relaxation of first and second layers often affects the layers further into the surface, for example by causing an expansion between second and third layers (usually of the order one per cent of the expected interlayer distance).

2.6.1 Surface reconstruction

Although the simple relaxation of a surface usually leads to measurable changes in interatomic distance at the surface, relaxation can also, in some cases, lead to a more substantial rearrangement of the surface atoms as they 'reconstruct' to form a structure where the atoms in the surface layer(s) occupy fundamentally different positions from those which would be expected if the bulk terminated abruptly at the surface.

This is particularly prevalent for semiconductor surfaces such as Si(111). Silicon has a diamond structure and on the close-packed (111) face, hybrid orbitals 'dangle' into the vacuum. This obviously creates a situation of high energy and so dramatic rearrangements are required to compensate for the loss of the nearest neighbours. Figure 2.27(a) shows the expected structure of the Si(111) surface, should no reconstruction occur. Each of the close-packed Si atoms in the diamond structure shown in the figure would have a dangling bond perpendicular to the surface. Figure 2.27(b) shows the reconstruction which occurs on this surface, the Si(111) 7 × 7. In the 7 × 7 reconstruction, the four top layers of the solid are altered. Within the parallelogram marked, 12 pyramids of Si are formed consisting of three atoms in the second layer with a single atom on top. Each of the 12 topmost atoms has a dangling bond which protrudes into the vacuum. Each pyramid surrounds another second-layer 'rest' atom, with the exception of the region in the central 'trough'. The six rest atoms also have dangling bonds. In the third layer the atoms adopt the usual surface structure apart from small shifts in interatomic distance, and some of the atoms are exposed to the vacuum as shown, including the central row separating two sets of six of the pyramids. Finally, within the parallelogram which has sides seven atoms long, the corner atoms are missing until the fourth layer down is reached. This elaborate reconstruction enables the Si(111) surface to adopt the minimum energy condition and effectively the highly unstable position of 49 dangling bonds is reduced to only 12 in the top layer and six in the second layer for the same unit cell.

Reconstructions are also observed for metal surfaces such as Au(110). The expected 'ploughed field' (110) structure is shown in Fig. 2.28(a) in both plan and profile forms. What is in fact observed for Au(110) is a reconstruction where a wider furrow is produced as shown in Fig. 2.28(b).

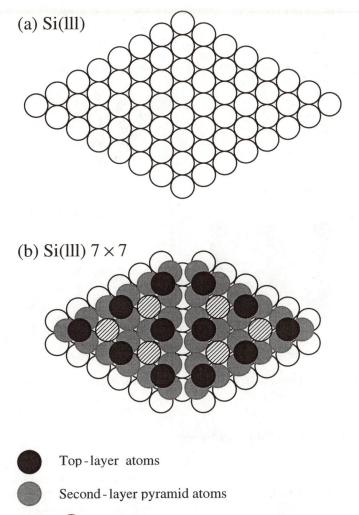

(a) Si(lll)

(b) Si(lll) 7×7

● Top-layer atoms

● Second-layer pyramid atoms

▨ Second-layer with dangling bonds

○ Third-layer atoms

Fig. 2.27 The Si(111) surface: (a) predicted on the basis of no reconstruction, and (b) the 7×7 reconstructed phase which is actually observed.

2.6.2 Techniques for studying surface structure

The most important methods used to identify surface structure, relaxations in surface structure and reconstructions have involved diffraction (of electrons and He atoms) and scanning tunnelling microscopy.

(a) Expected structure

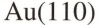

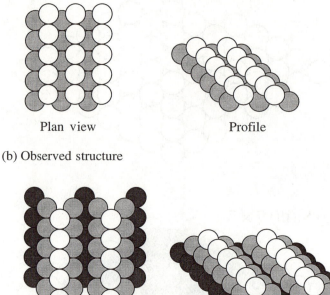

Plan view   Profile

(b) Observed structure

Plan view   Profile

Fig. 2.28 The Au(110) surface: (a) the expected structure, and (b) the observed reconstructed phase.

2.6.2.1 *Scanning tunnelling microscopy*

The development of scanning tunnelling microscopy (STM) has led to major breakthroughs in our understanding of surface phenomena because it is the only technique available which allows us to 'see' atoms by producing a 'real space' image of the surface. The technique was invented by Binning and Rohrer in the early 1980s. The principle is very simple but the experimental technique is highly demanding. An atomically sharp tip (ideally with only one atom at the tip) is brought up to a distance z ($\sim$0.5–1 nm) from a surface. If a voltage V ($\sim$10 V) is applied across the tip and surface, and if the distance z is small enough, a current i ($\sim$nA to pA) will flow between tip and surface. The current is able to flow because of overlap of the wavefunctions of the orbitals associated with the surface and the tip in a quantum mechanical tunnelling process. Obviously, as the distance between surface and tip is decreased, the orbital overlap will increase and thus the tunnelling current will increase.

If the voltage across surface/tip is held constant, then as the tip is moved across the surface, the STM operator can keep the current constant such that the tip follows the surface contours and a topographic image can be constructed by computer. The principle of the technique is shown in Fig. 2.29. The tip, to a first approximation, follows contours of a constant charge density in the surface. By applying a potential difference so that the surface is positive with respect to the tip, the filled states are

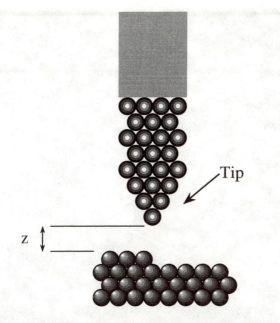

Fig. 2.29 The principle of the scanning tunnelling microscope. Under computer control the atomic tip is driven over the surface at a constant height, z, by maintaining a constant voltage and tunnelling current between the surface and tip.

studied, while a negative potential of the surface with respect to the tip gives an image due to the empty states.

A typical image is shown in Fig. 2.30; this is the STM image obtained for the Si(111) 7×7 which was shown in Fig. 2.27. The 'height' is indicated by shading of grey scale from dark (lowest) to light (highest). The parallelogram on Fig. 2.30 indicates the unit cell, which is shown schematically in Fig. 2.27(b). The 12 atoms that sit at the top of the three atom pyramids are in the lightest shades and the corner atoms which are four layers down are the darkest features shown, so the STM image is consistent with the description of the reconstruction which was given in Section 2.6.1.

Obviously, scanning an atomic tip over a surface at a height of < 1 nm to 0.01 nm is difficult experimentally, not least because of mechanical vibration. There are several measures which can be taken to reduce these problems; for example, tiny but massive STM systems can be adopted to damp out a lot of vibration (the natural frequency of something like this is very high); the entire vacuum system can be mounted on anti-vibration mounts to reduce coupling through the floor; and the systems can be (and usually are) constructed inside soundproofed rooms.

A further problem exists in that it is very difficult to obtain images of metallic substrates. It is usually far simpler to obtain an image of a semiconductor substrate than of a metal, because in a semiconductor the electron density is usually localized in specific directions, so the surface contours are quite corrugated. In a metal, the electron density is delocalized in the valence and/or conduction bands, especially in free-electron-like metals. This is slightly less true for transition metals, where localized

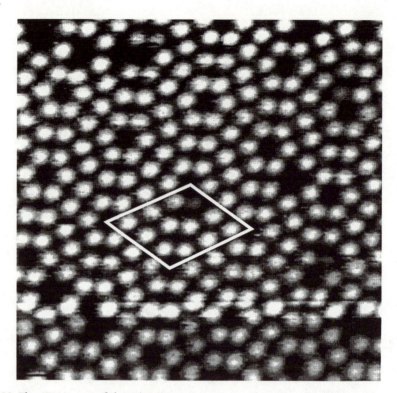

Fig. 2.30 The STM image of the Si(111) 7 × 7 structure. The unit cell which was shown in Fig. 2.27 (b) is indicated by the parallelogram. In the unit cell, the light circles are due to the 12 top-layer atoms and the corners appear dark because the atoms are in the fourth layer down. In other areas of the image, dark patches occur in unexpected positions and these indicate point defects and/or impurities. (Image recorded by Dr S.P. Tear of the Department of Physics, University of York.)

d electrons occur in the conduction bands. The image of a delocalized electron density is of an atomically flat region, like a plain sheet. STM can image metallic surfaces but it is usual to cool the substrate and the microscope down, using liquid helium, to obtain temperatures in the region 20–40 K to reduce thermal motion. Imaging of these surfaces also imposes more rigorous requirements on the need for isolation of the system.

It is important to realize that STM cannot be used to accurately determine interatomic distances between top and underlying layers; height information is strongly dependent on the electronic structure of the surface and this is subject to change, especially at step edges and kinks (for example). Even lateral, in-plane distances are not accurate and are subject to thermal drift (diffraction techniques such as those described in Section 2.6.2.2 are required for accurate measurements of lateral dimensions). In addition, STM cannot identify the chemical nature of atomic or molecular species bonded to the surface. The importance of STM lies in its use to observe the structure of the surface.

STM technology continues to improve and one of the current key areas is to use it to obtain spectroscopic information. If the tip is held in one position above the surface, the tunnelling current can be measured as a function of the voltage between the tip and the surface. The data obtained can be used to find the densities of occupied and unoccupied states and the technique is called 'inelastic tunnelling spectroscopy'. If the surface is held at a positive potential with respect to the tip, electrons tunnel out of the tip. The resulting current depends on the density of the occupied states in the tip, the density of empty states in the surface and the probability of tunnelling across the vacuum gap, z. If the tip is positive with respect to the surface, electrons tunnel in the opposite direction and the current depends on the density of the unoccupied states at the surface, the occupied states at the tip and the probability of tunnelling across z.

The method can be extended further by recording the variation of i with z for a number of points across the surface—a technique called 'current imaging tunnelling spectroscopy'.

Atomic 'nanoscale' manipulation is also of growing interest, where the aim is to move atoms and place them in desired positions on the surface. This can be done by pushing atoms around the surface with the tip or picking them up with the tip and placing them on the surface in the required position.

2.6.2.2 Diffraction methods

Until the advent of STM, diffraction methods dominated the study of the structure of surfaces. In fact, STM methods have not replaced diffraction because they are not sufficiently accurate in the determination of quantitative parameters such as interlayer and interatomic distances, and so diffraction methods are still important tools for gaining quantitative structural information.

The principle of diffraction is the same for the scattering of both electrons and atoms (usually helium atoms). One of the most commonly used diffraction techniques in surface science—low-energy electron diffraction—is described in detail below. The diffraction of helium atoms is also discussed; it is easier to interpret but more difficult experimentally and more expensive.

(i) Low-energy electron diffraction (LEED)

Diffraction methods rely on the use of the elastic scattering of monochromatic beams of incident energy, be they photons, atoms or, in this case, electrons. The use of particles to obtain diffraction information arises from the phenomenon of wave–particle duality. It is the wave-like properties that are important for producing diffraction features. The de Broglie equation is given in eqn (2.30) and relates the wavelength of a particle, λ, to its mass, m, and velocity, v.

$$\lambda = \frac{h}{mv} = \frac{h}{p} \tag{2.30}$$

where h is Planck's constant and p is the particle's momentum.

Bearing in mind that we are considering electrons in this case, we can calculate the effective wavelength in terms of the charge on the electron, and the voltage, V, through which the electrons (with charge e) are accelerated. The kinetic energy of the electrons E is given by

$$E = \frac{1}{2}mv^2 = eV \tag{2.31}$$

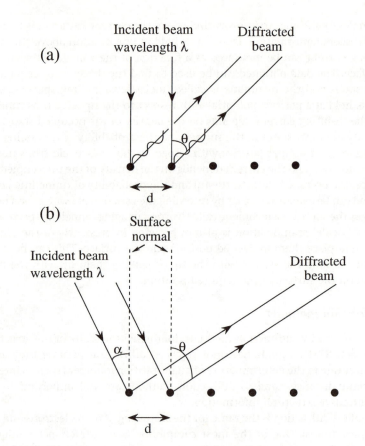

Fig. 2.31 Diffraction from a 1-d array of metal ion cores (a) with the incident beam perpendicular to the surface and (b) with the incident beam away from the surface normal.

so λ can be rewritten as

$$1 = \frac{h}{\sqrt{(2meV)}} = \left(\frac{150}{V}\right)^{1/2} \tag{2.32}$$

where λ is measured in angstroms and V in volts. To get reasonable diffraction, the wavelength of the electrons must be less than, but of roughly the same order of magnitude as, the spacings of the ion cores.

A second requirement of the electrons to get surface-sensitive information is that they must not penetrate too far into the sample; in fact they should only diffract from the top few layers.

Electrons which have energies in the range 20–500 eV are of the right order of magnitude to meet both requirements (see Fig. 2.20).

The diffraction effects can be explained by considering Fig. 2.31. An incident beam is scattered off a one-dimensional array of metal ion cores, and the scattered beams can interfere constructively or destructively where the path difference between electrons scattered from adjacent atoms is either $n\lambda$ or $n\lambda/2$, respectively. This is Bragg scattering, and eqn (2.33) can be applied:

$$n\lambda = d \sin \theta \tag{2.33}$$

where $n = 0$ or an integer for constructive interference. Here, d is the interatomic distance and θ the angle between incident and scattered beams, as shown in Fig. 2.31(a). For an incident beam which is not perpendicular to the surface, the expression is modified to

$$n\lambda = d(\sin \theta - \sin \alpha) \tag{2.34}$$

The diffraction observed arises from scattering from the atoms, but eqn (2.34) applies to scattering by all atoms lying within the incident beam and lying in rows of the same spacing, d. This means that each of the spots in the diffraction pattern arise from rows of atoms in the surface and can be indexed according to the two-dimensional Miller indices of the rows. Thus eqn (2.34) can be rewritten to include an accurate description of the Miller indices of the rows from which the diffraction is occurring.

$$n\lambda = d_{hk}(\sin \theta_{hk} - \sin \alpha) \tag{2.35}$$

where h and k are Miller indices.

A typical LEED system (a so-called retarding field analyser, RFA), is shown in Fig. 2.32. The sample is set at earth potential and a beam of electrons at energy E_p is fired at it so that the diffracted beam travels back towards the grids. The first grid the diffracted

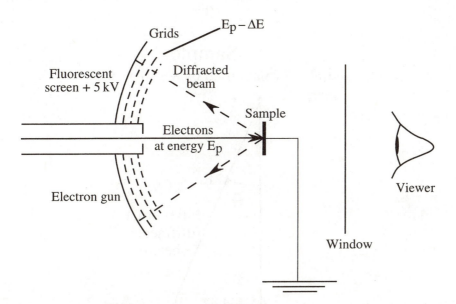

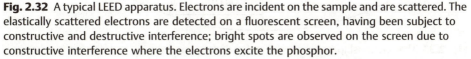

Fig. 2.32 A typical LEED apparatus. Electrons are incident on the sample and are scattered. The elastically scattered electrons are detected on a fluorescent screen, having been subject to constructive and destructive interference; bright spots are observed on the screen due to constructive interference where the electrons excite the phosphor.

beam meets is set at earth potential so that the electrons travel through a field-free region. The second grid is set at a slightly less negative bias than the primary beam energy to remove any inelastically scattered electrons. The final grid is set at earth potential to reduce the effect of the field caused by the screen voltage on the second grid. The fluorescent screen is set at a large positive potential to accelerate the electrons so that they excite the phosphor coating on impact. The fluorescent spots are produced on the screen where constructive interference occurs and are observed via the view port outside the uhv chamber. An image of the LEED pattern is usually recorded by using a camera with high-speed film or a digital camera attached to a computer.

A typical LEED pattern is composed of a set of diffraction spots which looks like a lattice. The lattice of spots is directly related to what is called the reciprocal of the surface structure. We can relate the LEED pattern, which can be described as the reciprocal mesh or lattice, to the 'real space' mesh of the surface itself. It should be noted that the LEED pattern may show only some of the reciprocal net points because of systematic absences (these are expected spots that are consistently missing from the pattern) or very low intensities.

We can find the unit mesh in real space by looking at the reciprocal lattice and relating the reciprocal dimensions and patterns observed to the real space dimensions and structure. This is done on the basis of simple geometry, as shown in Fig. 2.33.

A much enlarged 1-d array is indicated separated by distance d_{10} as shown; the Bragg equation applies and for first-order diffraction can be written

$$\sin \theta = \frac{\lambda}{d_{10}} \tag{2.36}$$

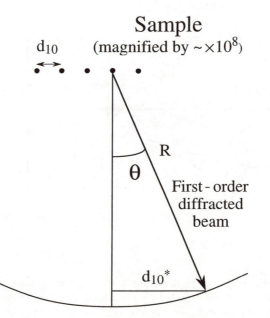

Fig. 2.33 The relationship between real and reciprocal meshes; d_{10} and d_{10^*} are indicated.

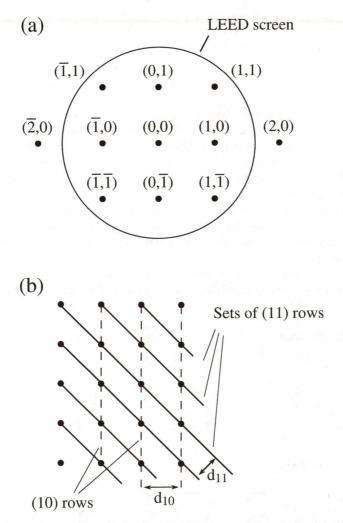

Fig. 2.34 (a) The reciprocal mesh observed by LEED for a (100) surface, with the indices to describe the observed LEED spots. (b) The (11) and (10) rows in real space which result in the (11) and (10) spots on the LEED pattern. It should be noted that $d_{11}/d_{10} = \sqrt{2}$.

From this equation, a value for the interatomic spacing can be deduced, using θ, which can be measured easily from the LEED pattern. The simplest method is to place the sample at the centre of curvature of the LEED screen. The distance of the first-order diffraction spot from the undiffracted (00) spot, d_{10}^*, can be measured and, with the radius of the LEED system, R, the scattering angle subtended at the sample, θ, can be deduced from

$$\sin \theta = \frac{d_{10}^*}{R} \tag{2.37}$$

Because the wavelength of the electrons is known, eqns (2.36) and (2.37) can be combined with the radius of the LEED system so that a measurement of d_{10}^* from the screen allows us to calculate the interatomic spacing of the rows,

$$d_{10} = \frac{R\lambda}{d_{10}^*} \tag{2.38}$$

If we now consider the pattern which we see with LEED more closely, we can describe the origin of each of the spots in terms of the 2-d Miller indices of the rows from which they result. Figure 2.34(a) shows the LEED pattern which would result from a square lattice; each of the spots is indexed. The (0,0) spot is the result of zero diffraction while those at (1,0) and (1,1) result from scattering from adjacent rows in the (10) and (11) directions as indicated in Fig. 2.34(b).

It should be noted that the higher the energy of the beam (the shorter λ), the closer together the spots appear on the screen. When carrying out LEED experiments it is usual to record patterns at two or more beam energies to ensure that all features of the pattern are clear.

Figure 2.35 shows the real and reciprocal space meshes expected for the low indexed faces of an fcc sample. Our original description of real and reciprocal meshes was given

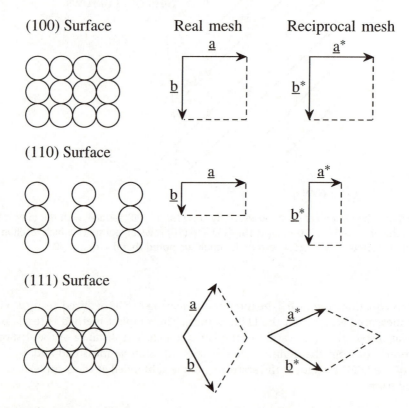

Fig. 2.35 Plan views, real and reciprocal space meshes for three simple fcc structures: (100), (110) and (111) faces.

in terms of 1-d structures; however, from the above it is clear that each spot results from adjacent rows of atoms and the reciprocal and real space meshes are in fact described by 2-d vectors. For the (100) surface, both vectors $\underline{a}$ and $\underline{b}$ are of the same length and so are obviously of the same length in reciprocal space. For the (110) surface $\underline{a}$ is longer than $\underline{b}$ and so in reciprocal space the lengths appear to be the other way round, i.e. $\underline{b}^* > \underline{a}^*$. For the (111) surface there is an added complication in that taking the reciprocal of the real space lattice changes the angle between the vectors from 120° in real space to 60° in reciprocal space.

Another way of describing diffraction is in terms of electron wavevectors and reciprocal lattice vectors. The magnitude of the incident wavevector of an electron $|\,\underline{k}\,|$ is given by $|\,\underline{k}\,| = 2\pi/\lambda$ (eqn 2.16). This is a measure of the electron's momentum and is deduced from eqn (2.30).

The reconstruction of a surface, or binding of molecules to the surface (discussed in Chapter 3), causes additional spots to appear in the pattern, and their position and intensity can be used to deduce the surface structure.

(ii) Helium atom scattering (HAS)

A full structural analysis of a surface using a diffraction technique requires measurement of the intensities of the diffraction spots and their positions. This is especially true when the surface is composed of more than one type of atom, or has reconstructed. It may be that when a different type of atom bonds to the surface, the same diffraction pattern results as for the clean surface, for example if atoms sit directly above the underlying substrate atoms. The LEED spots appear in the same positions; the only thing that changes in this situation is their intensity. The intensities observed are dependent on the strength of scattering from each type of atom on the surface, and there may also be changes which are induced because of changes of phase between one type of atom and another, causing changes in the amounts of constructive and destructive interference. In order to undertake a full analysis, it is therefore necessary to do a complicated intensity analysis to get the full surface structure.

For LEED, a full intensity analysis is not a routine matter (as it would be in the 3-d X-ray structural analysis of solids), because of the strength of scattering of electrons. X-rays can penetrate deeply into the solid without interacting substantially with the atoms in it, and so are treated with so-called 'kinematic scattering' theories, which consider a direct and uncomplicated scattering mechanism. Electrons, on the other hand, interact strongly with the atoms in the array; they suffer from multiple diffraction events and from complications arising from phase variations (in their wave properties), which means that they have to be studied by so-called 'dynamical scattering' theories.

The use of helium atom scattering (HAS) overcomes the need for the use of these complex dynamical scattering theories, as the interaction of helium atoms with a surface is well understood and relatively simple. This is mainly because, unlike electrons, the helium atoms only interact with atoms in the surface itself rather than interacting with surface atoms and those in the layers just beneath it. The far simpler kinematic theories can therefore be applied to HAS data to determine surface structure. The elastic scattering of helium atoms results in the observation of diffraction patterns caused by constructive and destructive interference in the same way as they are produced for LEED. Unfortunately, HAS is far more difficult to perform technically and is very expensive (the apparatus required for HAS is described in Chapter 4, Section 4.6.2). It is also necessary to have surfaces which are highly ordered; usually, surfaces must be ordered to a far greater degree than those from which LEED patterns can be obtained. This is a result of the 'coherence' properties of the incident beams of He atoms and electrons. A helium beam is coherent (i.e. has the same phase) over a wider area than a typical electron beam, which means that the diffraction features observed for HAS arise from a larger area of surface than those obtained in LEED. Any disorder in this area of surface causes destructive interference. The effect is greater on HAS compared to LEED simply because of the larger total amount of disorder within the area being sampled.

However, despite these problems this technique has proved very informative for elucidating several important problems in surface science.

(iii) 'Fitting' data to obtain surface structures

The usual way in which the deduction of a structure is carried out is by an iterative fitting process. Typically, an educated guess at the real space structure is made and this is fed into a computer package that is designed for the task. The program then calculates the reciprocal mesh and the intensity pattern that would result from the proposed structure. The generated reciprocal pattern is compared with the observed pattern, usually using some kind of 'least-mean-squares' fitting procedure. If the data are statistically 'close' to the suggested structure, it is assumed to be the correct one. If the data are not 'close' enough, a new 'refined' structure is suggested. The reciprocal mesh and its intensity are then generated for the new structure and this is compared again to the observed structure. The whole process continues until the 'fit' is judged to be statistically close enough to describe the structure.

The computer packages used encompass the 'best' type of scattering theory for the technique used. In the case of HAS, a kinematic package is used, while for LEED, packages with time-consuming and complex dynamic theory are used.

3
Adsorption and desorption

3.1 Adsorption processes

3.1.1 Atomic/molecular collisions with surfaces

There are several possible outcomes when an atom or a molecule hits a surface. The first is that the atom or molecule merely bounces back off the surface. This is called 'elastic scattering', as in this mechanism there is no energy exchange between the surface and the incident atom or molecule. This is essentially the process that occurs when helium atoms are scattered from surfaces in HAS experiments, as described in Chapter 2.

A second possibility is that as an atom or a molecule rebounds from the surface, it loses or gains energy in an 'inelastic scattering' process. This can be likened to the Raman scattering process where the incident photons lose or gain quanta of vibrational and/or rotational energy.

Both of these outcomes can be used to our advantage in the development of techniques for studying surface phenomena (such as HAS), but the outcome of an atomic or a molecular collision that results in the retention of the molecule on the surface is of far greater importance to our study of surface chemistry. There are two types of interactions that can occur. In broad terms these are physical adsorption (or physisorption) and a much stronger interaction, which results in the formation of one or more chemical bonds between the adsorbed molecule and the surface (chemisorption). In each case the atom or molecule being adsorbed on the surface is usually described as the adsorbate; the adsorbing surface is usually termed the adsorbent or substrate.

3.1.2 Coverage

The coverage, θ', of adsorbate on the surface is defined as

$$\theta' = N_{ads}/N_s \tag{3.1}$$

where N_{ads} is the number of adsorbate atoms per unit area and N_s is the number of surface atoms per unit area. This expression can be used to compare the number of atoms or molecules directly adsorbed on the surface in the first layer with the number of atoms in the surface. Another way of describing the amount of adsorbate on the surface is by using the fractional coverage, θ, which is given

$$\theta = N_{ads}/N_m \tag{3.2}$$

where N_m is the number of atoms/molecules adsorbed per unit area to produce one complete layer on the surface. The maximum amount of adsorbate which is adsorbed in the first layer is usually called a monolayer, so $\theta = 1$; θ', on the other hand, is commonly less than 1, as for example, when the adsorbed molecule is larger than the substrate atoms. (On some occasions the coverage θ' can be found to be greater than 1; for adsorption of H on W, for example, it is 2, because two H atoms can be adsorbed on each W atom.)

3.1.3 Physisorption

Physisorption involves the balancing of a weak attractive force, for example, of a Van der Waals nature, between the surface and the adsorbate, with the repulsive force associated with close contact. The process is always exothermic and the energy given out on adsorption, *the heat of adsorption*, ΔH_{ads}, is low, typically in the region -10 to -40 kJ mol^{-1}. In these systems the adsorbate–adsorbate interactions are often comparable with, and can be considerably greater than, the adsorbate–surface interactions. (This will be dealt with in greater detail in Chapter 5).

A Van der Waals interaction can be described approximately by the Lennard-Jones (12–6) potential as for molecule–molecule interactions, for which the potential energy $V(r)$, is given by

$$V(r) = 4\varepsilon \left[\left(\frac{\sigma}{r}\right)^{12} - \left(\frac{\sigma}{r}\right)^{6} \right] \tag{3.3}$$

where r is the interatomic distance, ε is the well-depth of the potential energy curve and σ is the interatomic distance at which the potential energy curve crosses zero. The origin of the parameters in eqn (3.3) comes from the attractive interaction between atoms/molecules which has the $1/r^6$ dependence and the repulsive interaction which has a $1/r^{12}$ dependence.

The repulsive forces become significant as the atoms/molecules approach each other closely. At the equilibrium bond distance the repulsion between electron clouds just balances the attractive forces and at this point the curve crosses the zero in energy $V(r)$.

The attractive forces are due to a combination of dispersion and dipolar forces. For a molecule approaching a surface, dispersion forces arise from instantaneous fluctuations in electron density which cause transient dipoles in the molecule; these interact with their opposite numbers in the surface atoms. Forces due to molecules that have permanent dipole moments are usually stronger. These permanent dipoles interact with electrical asymmetries in the surface. The presence of permanent dipoles can also induce dipoles in polarizable non-polar molecules and in polarizable surfaces.

The situation when a molecule approaches a surface is more complex than that which occurs when two atoms/molecules approach each other. The potential energy can be calculated as a function of distance between surface and adsorbing molecule by a pairwise summation of all interactions between the incoming molecule and the

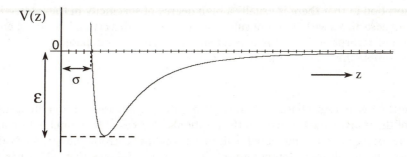

Fig. 3.1 The 9–3 potential well for physisorption. σ indicates the interatomic distance at which the potential energy curve crosses zero.

nearest solid atoms. The incoming molecule interacts with each of the atoms in the solid and each of these interactions can be described by a Lennard-Jones (12–6) pair potential. Summing the pair potentials of the adsorbing molecule with each atom in the solid leads to an expression of the form given in eqn (3.4), where the potential energy is given as a function of the height of the molecule above the surface, z.

$$V(z) = 4\pi\varepsilon N_s\sigma^3\left[\frac{1}{45}\left(\frac{\sigma}{z}\right)^9 - \frac{1}{6}\left(\frac{\sigma}{z}\right)^3\right] \tag{3.4}$$

For physisorption, the strength of interaction is greater than that which is found in other weakly interacting systems because the interaction with each volume element in the solid is taken into account. This means that a (9–3) potential applies rather than a (12–6) potential. The (9–3) potential is plotted in Fig. 3.1, with the parameters used indicated on the figure. The molecule adopts an equilibrium height above the surface which corresponds to the minimum of the potential well, ε, and this well depth is the heat of adsorption, ΔH_{ads}. ΔH_{ads} is sometimes called q, where $q = -\Delta H_{ads}$. This nomenclature was adopted for convenience because q is always positive but it is slowly falling into disuse.

Generally speaking, physisorption is non-specific and any atom or molecule can adsorb on any surface under appropriate experimental conditions (i.e. temperature and pressure); large amounts of physisorption are favoured when the surface is at low temperature. The heat of adsorption is usually of the same order of magnitude but slightly larger than the latent heat of condensation of the adsorbate molecules.

3.1.4 Chemisorption

Chemisorption describes the process by which a strong chemical bond is formed between the surface and the adsorbate. It involves the exchange of electrons between the adsorbing molecule and the surface. The heats of adsorption for chemisorption are generally larger than for physisorption; some are very large. ΔH_{ads} tends to be in the range -40 to -1000 kJ mol^{-1} and it is found that chemisorbed layers with large heats of adsorption tend to be very stable to high temperatures. The bonds formed may be ionic or covalent, or a mixture of the two. Chemisorption differs markedly from physisorption in that there is usually a high degree of specificity in the interaction of different adsorbates with different substrate surfaces. With a crystalline adsorbent, the strength of adsorption is usually dependent on which face is exposed.

From simple thermodynamic considerations,

$$\Delta G = \Delta H - T \Delta S \tag{3.5}$$

so that the free energy change ΔG must be negative. Upon adsorption onto a surface, most of the translational, and often the rotational, degrees of freedom are lost, and so the order of the system is increased. This means that the entropy change, ΔS, is always negative. As the absolute temperature T is positive, it follows that ΔH_{ads} has to be negative; thus, in order to proceed, the chemisorption process *must* be exothermic.

Several factors might cause a strong coverage dependence for ΔH_{ads}. For physisorbed species it can vary because of the effect of electrical polarization which changes as the adsorbates become more closely packed on the surface. In the case of chemisorbed species, the charge transfer between surface and adsorbate can cause later adsorption to be less energetically favourable. Additional effects can be due to electrostatic coupling between adsorbates or repulsive forces which occur when the chemisorbed species are packed closely on the surface. For this reason ΔH_{ads} is always quoted for a given coverage of adsorbate on the surface.

Chemisorption resulting in the formation of a chemical bond between the adsorbing molecule and the surface is termed non-dissociative chemisorption; this applies when all the original interatomic bonds of the adsorbed molecule are retained in some form. Adsorption can, alternatively, result in the dissociation of the molecule on binding to the surface. Chemisorption in this manner frequently requires activation. Chemisorption processes can be described by potential energy curves for non-dissociative and dissociative processes as detailed below.

It should be noted here that for chemisorption, while the bonding between adsorbate and substrate can be represented by the Lennard-Jones (12–6) potential, it can alternatively be represented by a Morse curve (as used in diatomic molecular spectroscopy).

3.1.4.1 *Non-dissociative chemisorption*

The non-dissociative chemisorption process for molecule AB can be written

$$AB \rightarrow AB_{ads} \tag{3.6}$$

The potential energy curve, which is based on a Lennard-Jones (12–6) potential, $V(z)$, is shown in Fig. 3.2. The distance between the adsorbing molecule and the surface, z, is presented on the x-axis. The potential for the weakly bound physisorption state is shown as a dashed line and that for the chemisorption well as a solid line. As the molecule approaches the surface it drops into the weakly bound (often short-lived) physisorbed precursor state, then passes over a small energy barrier and into the chemisorption well as shown by the solid curve in the figure.

An example of this sort of chemisorption is of CO adsorption on low indexed copper surfaces.

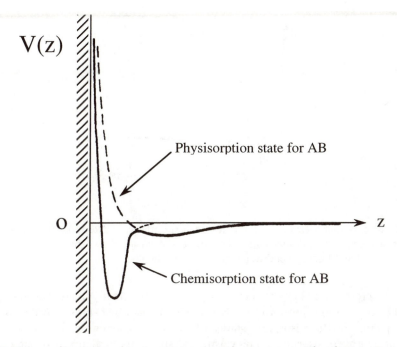

Fig. 3.2 The non-dissociative chemisorption potential energy curve. The physisorption state is shown by a dashed curve and the lowest energy pathway is shown as a solid line.

3.1.4.2 *Dissociative chemisorption*

As indicated above, there are two forms of dissociative chemisorption; activated processes requiring additional energy and non-activated which proceed because they are energetically favourable. In this case we will consider the simple dissociation

$$AB \rightarrow A_{ads} + B_{ads} \tag{3.7}$$

(i) Activated dissociative chemisorption
The potential energy curve for activated dissociative chemisorption is shown in Fig. 3.3. Here the curve with a dashed section shows the physisorption of AB on the surface. It crosses the dashed–dotted curve, which shows the interaction of the A + B fragments with the surface. The important point to note here is that the two curves cross above the zero of potential energy. The minimum energy pathway is shown as a solid curve. In the absence of additional energy the molecule is adsorbed in the physisorption well. The molecule can only cross over into the dissociated chemisorption state if additional energy is supplied to overcome the activation barrier, which is above the zero in potential energy.

(ii) Non-activated dissociative chemisorption
The potential energy curves for non-activated dissociative chemisorption are shown in Fig. 3.4. The curves representing physisorbed AB and chemisorbed A + B fragments are essentially the same as those for the activated process in Fig. 3.3, but in this case the

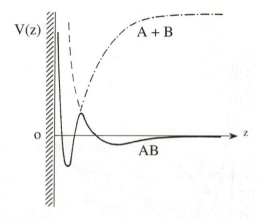

Fig. 3.3 The potential energy curve for activated dissociative chemisorption. The physisorption state for AB is shown by a dashed curve, the dashed–dotted line shows the adsorption state for A + B and the lowest energy pathway is shown as a solid line.

two curves cross just below the zero of potential energy. Thus, no additional energy is needed to pass the system into the dissociated chemisorbed state. In order to gain a mental picture of the adsorption states, idealized representations of the adsorption states for hydrogen adsorbing on nickel are also shown in the figure, associated with z. Initially a physisorbed precursor state is formed and the molecule then forms a higher energy transition state, at the top of the activation energy barrier E_a, before finally passing into the chemisorption potential well.

A good example of dissociative chemisorption is that of H_2 adsorption on a metal surface. When a diatomic molecule nears a surface, an equilibrium as given in eqn (3.8) is set up:

$$M + H_{2(g)} \leftrightarrow 2\,M–H_{ads} \tag{3.8}$$

The position of the equilibrium depends on the reaction conditions, particularly in terms of pressure and temperature. As we have seen earlier, ΔH_{ads} must be negative for the dissociation to proceed and this condition is met when $2 \times D_{M–H} > D_{H–H}$, where D denotes the dissociation energies of the M–H and H–H species, respectively. From eqn (3.5) it can be seen that when ΔH_{ads} is small, the position of the equilibrium is strongly dependent on $T\Delta S$. Whether or not the process is activated is governed by the precise energetics involved as the molecule approaches the surface.

A striking example of this is for the dissociative adsorption of molecular hydrogen on copper. At a pressure of 10^{-6} mbar and at room temperature, negligible quantities of atomic hydrogen are adsorbed. However, at the same temperature, a pressure of 100 mbar produces a monolayer of Cu–H species on the surface.

Dissociated species appear generally at higher temperatures than non-dissociated species because the system has to be carried over the activation energy barrier to dissociation for both activated and non-activated dissociative chemisorption.

The three situations are described here in only schematic, one-dimensional representations and are, in any case, very simple systems. In more complex systems

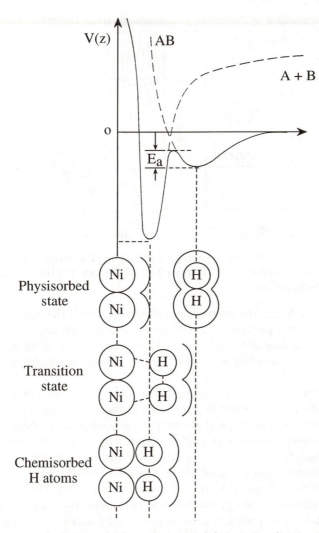

Fig. 3.4 The potential energy curve for non-activated dissociative chemisorption with the states which are formed for the adsorption of molecular hydrogen on nickel.

there are often several possible precursor states to chemisorption, and much closer similarity between energies of the non-dissociated and dissociated states which are in competition. This often makes the possibilities within the adsorption process more complicated.

3.1.5 Adsorption dynamics and the potential energy hypersurface

Figure 3.5(a) shows the conventional schematic diagram for the potential energy curves for the dissociative adsorption of molecule AB on a surface (from Fig. 3.3). The energy is drawn as a function of a one-dimensional pathway. As we noted in the

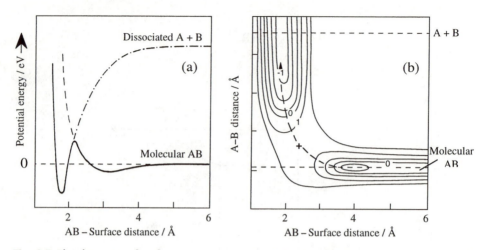

Fig. 3.5 The dynamics of surface processes. (a) the 1-d potential energy curve for activated dissociative adsorption (from Fig. 3.3) and (b) a 2-d representation of the same process, the potential energy hypersurface.

previous section, although the curve gives a simple picture of the energetics involved in adsorption, it is not particularly useful for obtaining anything more than a very qualitative understanding of the adsorption/desorption process, or for making overall predictions about the adsorption process. Such phenomena require more complex consideration.

Figure 3.5(b) shows a view of the same process, plotted as a function of two reaction parameters, which gives far better insights into the detail of the atomic motions of the surface and adsorbates, the so-called *dynamics* of the adsorption process. This form of plot is called a potential energy hypersurface and such plots are commonly produced to describe gas-phase chemical adsorption dynamics. The contours show lines of equal energy and the two dashed lines show the cuts across the potential energy hypersurface which correspond to the 1-d curves in Fig. 3.5(a). The arrows indicate the pathway of minimum energy between the two states. From the 2-d plot we can begin to understand how a concerted mechanism for dissociation is required; in other words how, as the AB bond weakens, the surface bond strengthens.

The translational and vibrational energies involved are very important and their effects are illustrated in Fig. 3.6 which shows the energies related to the molecular interatomic distance, a, and molecular height, z, from the surface. There are two characteristic reaction scenarios which are commonly found, shown in Fig. 3.6(a) and (b). The transition state is indicated by the 'seam' at the point at which it crosses the minimum energy pathway. In the first example, Fig. 3.6(a), this seam sits in the 'entrance channel' which describes the situation when the molecule is not stretched as it leaves the physisorption well and moves in towards the surface. Thus the bond length of the transition state species is not significantly different to that of the precursor. If the translational kinetic energy is increased at this stage, the rate of dissociation is increased.

The other scenario is shown in Fig. 3.6(b). In this case, the adsorbate is closer to the surface as it sits in a physisorption well for which the adsorbate–substrate distance is

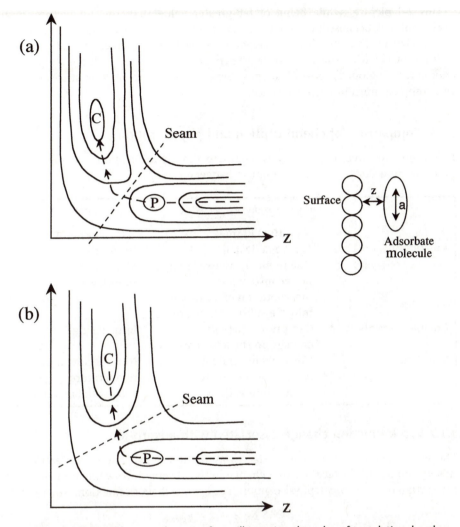

Fig. 3.6 The 2-d potential energy hypersurfaces illustrating the roles of translational and vibrational energy states in the two extreme cases of activated dissociative adsorption of a diatomic molecule: (a) where the 'seam' sits in the 'entrance channel' and (b) where the 'seam' sits in the 'exit channel'. The molecule has interatomic distance, a, and the contours are plotted as a function of z, the molecular distance from the surface.

close to that adopted by the molecular fragments once they have dissociated. Vibration of the physisorbed molecule (which causes an increase in the interatomic distance) then leads to dissociation and provides the rate-limiting barrier. The transition state is said to lie in the 'exit channel', and any increase in translational energy has little effect on the dissociation process. An increase in vibrational energy, on the other hand, promotes the dissociation (called 'vibrationally assisted sticking').

This 2-d picture, while being of far greater value than the 1-d picture, is also oversimplified, because it does not depict enough degrees of freedom. Even in the case of dissociation of a diatomic molecule there are six degrees of freedom for the molecule as it interacts with a static substrate. Effectively, what we see with the 2-d plot is the chosen freezing out of most of the possible coordinates, as we look at slices through this complex multidimensional space.

3.1.6 Comparison of chemisorption and physisorption

It is useful to have a summary to compare the chemisorption and physisorption processes in order to help to distinguish between them.

	Chemisorption	*Physisorption*
$-\Delta H_{ads}$	~40–1000 kJ mol^{-1}	~10–40 kJ mol^{-1}
Kinetics of activation	Can be activated	Non-activated
Number of layers	One monolayer (assuming no reconstruction or incorporation of adsorbate into the subsurface region)	Monolayers and multilayers
Chemical reactivity	Can cause reactivity changes in the adsorbate	Little change
Specificity	Normally dependent on specific adsorbate–surface interactions	Non-specific, needs low temperature to get substantial amounts

3.1.7 Work function change associated with adsorption

The work function of a clean surface was discussed in Chapter 2, Section 2.2.2. Adsorption on the surface causes a change in the observed work function, even if the interaction is that of weak physisorption. The change in work function, $\Delta\phi$, is given by

$$\Delta\phi = \phi_{ads} - \phi_{clean} \tag{3.9}$$

where ϕ_{ads} is the work function of the adsorbate-covered surface at a given coverage θ or θ', and ϕ_{clean} is the work function of the clean surface. The change in the work function is a valuable parameter for assessing the degree of charge redistribution following adsorption.

The distribution of the charge on the surface or surface dipole has considerable influence on the atoms or molecules that adsorb. It can affect the molecular shape and orientation adopted on the surface and can lead to fragmentation of the adsorbate. The field at the surface even affects inert gas atoms such as Ar and Xe as they are polarizable, and so dipoles are produced within the atoms themselves during adsorption.

The effect of inert gas adsorption on the work function is not what might be expected from a naive, classical prediction. In Section 2.2.1 it was seen that the electron density of the clean surface spills out towards the vacuum. In the classical picture, the electrons of the incoming atoms would be repelled by the surface such that a dipole would be set up within them, with the positive end closest to the surface. This would cause an increase in the work function because the dipoles set up in the atoms would add to the charge at the surface.

However, the observed effect is that a decrease in work function occurs, which shows that the inert gas atoms are polarized in the opposite direction, where a negative charge builds up between the atoms and the surface. The reason is concerned with a subtle effect due to a quantum mechanical electron correlation effect. This arises because the electrons can attain lower energies by behaving cooperatively with the free electrons in the metal; they therefore have a preference to move to a region of higher electron density. This phenomenon cannot be understood classically.

The resulting arrangement of adsorbed atoms and their dipoles is shown schematically in Fig. 3.7 for physisorbed Ar on a metal surface.

The situation can be modelled as a parallel plate condenser and so the Helmholtz equation can be applied. This is given in eqn (3.10) where ΔV is the change in surface potential, μ is the induced dipole moment, which is taken by convention to be positive in the $+ \rightarrow -$ direction, and ε_0 is the permittivity of free space ($\sim 8.85 \times 10^{-12}$ C V^{-1} m^{-1}):

$$\Delta V = \frac{N_{ads}\,\mu}{\varepsilon_0} \tag{3.10}$$

The change in surface potential is related to the work-function change by

$$\Delta \phi = \Delta V\, \mathrm{e} \tag{3.11}$$

where e is the charge on the electron. A measurement of ΔV or $\Delta \phi$ can therefore be used to give an estimate of the partial charge on the adsorbate, Q, because

$$\mu = Q\, a \tag{3.12}$$

where a is the separation of the adsorbate and the screening charge.

Chemisorption often causes large changes in the observed work function. This is because there is usually a high degree of charge transfer between the adsorbate and the surface when strong chemical bonds are formed. Intuitively, one can predict that adsorbates which are electronegative with respect to the surface will have the opposite effect on ϕ compared to the electropositive ones. The way in which they affect ϕ depends also on the 'double charge layer' which is induced by the surface itself (as discussed in Section 2.2.2). Figure 3.8 shows a schematic energy level diagram illustrating the two situations. The highest occupied molecular orbital (HOMO) for the adsorbing molecule is depicted in the figure, with the ionization energy of the level indicated by I.

(i) Electropositive adsorbates
Figure 3.8(a) depicts the situation for electropositive adsorbates which donate electrons towards the surface. When the first ionization energy of the adsorbate, I, is lower than the work function of the surface, electron transfer to the surface takes place; this transfer ceases when the Fermi level and the HOMO are equal. The electron transfer to

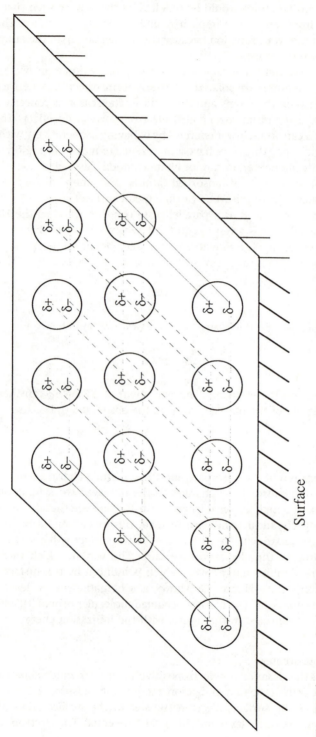

Surface

Fig. 3.7 The array of induced dipoles for argon atoms physisorbed on a metal surface. The dipoles act as a parallel plate condenser.

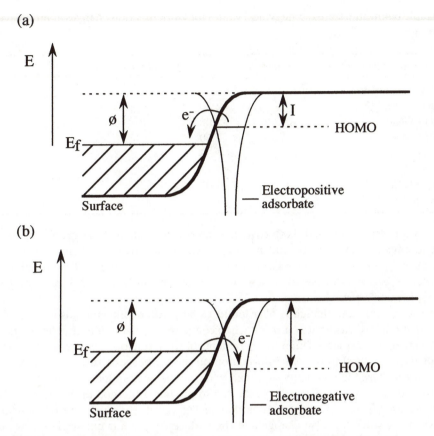

Fig. 3.8 Energy level diagrams showing the effect on the work function of (a) an electropositive adsorbate and (b) an electronegative adsorbate. The Fermi level, E_f, is indicated for the surface and the highest occupied molecular orbital (HOMO) of the adsorbate is indicated with I, the first ionisation energy of the level.

the surface results in a positive charge on the adsorbates. The resulting effective dipoles are in the opposite direction to those of the clean surface and so these counteract the double layer due to the surface itself and produce a *reduction* in the work function. It should be noted that for conducting surfaces such as metals, image charges are set up within the surface which enhance the dipoles due to the adsorbates.

(ii) Electronegative adsorbates
Figure 3.8(b) depicts the situation for electronegative adsorbates, where the ionization energy, I, of the adsorbate is greater than ϕ. The electronegative adsorbate withdraws electron density from the surface, resulting in a negatively charged adsorbate. The dipole that is produced is in the same direction as that of the clean surface and so enhances the double layer due to the surface itself, and *increases* the work function. Again, for conducting surfaces, the dipole produced by the adsorbate is enhanced because of the image charge produced in the surface.

The table below illustrates some typical values for work function changes at monolayer coverages.

Typical work function changes for monolayers ($\theta = 1$) of adsorbates	
Substrate/adsorbate	$\Delta\phi$/eV
Rh(111)/Na	−2.30
Ni(110)/Na	−2.00
Mo(100)/K	−2.00
Rh/C_2D_6	−1.36
Rh(111)/CCH_3	−1.23
Rh(111)/CO	+1.05
Pd(100)/H_2	+1.20

The work function change is both coverage and site dependent. A typical plot of work function against coverage dependence for an electronegative adsorbate is shown in Fig. 3.9(a). The coverage dependence arises from the total amount of charge that is transferred between the adsorbate and the surface. Obviously at lower coverages, less charge is transferred.

The work function change is also found to be greater following adsorption of a known amount of adsorbate on steps, when compared to that of identical amounts of adsorbate on terrace sites. Often measurements of work function change as a function of coverage are used to provide information on the surface structure and step/defect densities as these higher energy sites tend to be populated first. Indeed, work function measurements can sometimes be calibrated to provide a method of monitoring the surface coverage, using plots such as that in Fig. 3.9(a). However, the relationship between the work function change and the surface coverage is not straightforward, because adsorbate–adsorbate interactions are also involved. For example, the initial adsorption of alkali metals onto metal surfaces produces a rapid decrease in work function as shown in Fig. 3.9(b) in line with the discussion in Section 3.1.7. The linear decrease is due to the increasing total amount of negative charge transfer with increasing coverage. At a certain coverage of alkali metal, however, the now partially positively charged adsorbates are forced closer together and repel each other. The repulsion effectively 'depolarizes' the adsorbate–surface dipoles and so reduces the overall charge transfer between the alkali metal and the surface. The work function therefore increases again and eventually reaches a plateau, at which point the alkali atoms have formed a metallic overlayer.

3.1.8 The Langmuir

At this point it is useful to define quantities for the amounts which are dosed onto the surface. When a surface is exposed to a gas, the dosage is usually recorded in terms of the pressure of adsorbing gas and the length of time it is present in the chamber, as these factors determine the number of molecular collisions with the surface. In Section 2.5.1 it was stated that at 'normal' temperatures (e.g. 300 K) a reactive gas of moderate molecular mass would completely cover a surface in one second if it were at a pressure

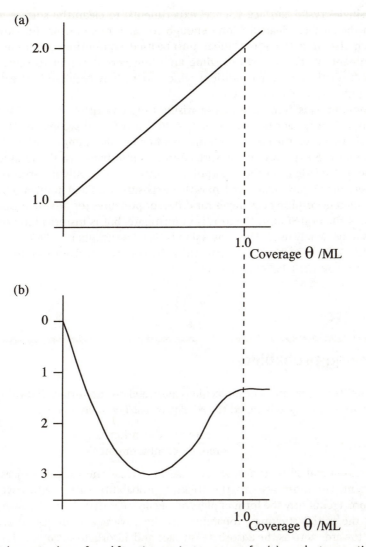

Fig. 3.9 Schematic plots of workfunction against coverage for (a) an electronegative adsorbate on a surface and (b) an adsorbed alkali metal.

of $\sim 10^{-6}$ mbar (10^{-4} N m^{-2}) and has unit sticking probability (see Section 3.2). This can be calculated using simple gas-phase collision theory; the molecular collision rate per unit area is given by

$$\mathrm{d}N/\mathrm{d}t = P/(2\pi m k T)^{\frac{1}{2}} \tag{3.13}$$

where N is the number of molecules, P the gas pressure, m the molecular mass, k the Boltzmann constant and T the temperature. The rapid rate at which the surface is covered by reactive species illustrates why ultrahigh vacuum conditions are usually used for single-crystal 'surface science' experiments; to clean the surface atomically and keep the surface clean for long enough to carry out meaningful experiments, pressures in the range 10^{-9}–10^{-10} mbar must be used. Equilibrium conditions obtained for experiments in this pressure regime are often very different to those obtained under the 'realistic' pressure conditions which are usually required for 'real' catalytic reactions (often several atmospheres).

The Langmuir (L) is defined as a dose arising from an exposure of 10^{-6} torr for one second; this is equivalent to an exposure of 10^{-7} torr for 10 seconds or 10^{-8} torr for 100 seconds. However, the Langmuir term can often be misleading as although it gives a total value for the exposure of the surface to the adsorbate gas, the total amount of adsorption is dependent on the adsorption conditions, especially in terms of pressure and temperature. A particular problem is that exposure to a fixed number of Langmuirs can give different resulting coverages for different pressures (or fluxes) of gas.

The use of the non-SI unit, the torr, is unfortunate, but is historical in origin and it continues to be widely used. The conversion is that 1 atmosphere = 760 torr = 1013.25 mbar = 1.01325×10^5 N m^{-2} (or Pa). Thus the conversion from torr to mbar is by multiplication by $1013.25/760 = 1.333$.

3.2 Sticking probability

The probability that an atom or molecule is adsorbed on the surface following a single collision with the surface is called the *sticking probability*, s, defined by

$$s = \frac{\text{rate of adsorption}}{\text{rate of bombardment}} \tag{3.14}$$

For many gas–metal systems at low coverages, $s \sim 1$, i.e. the sticking process is often very efficient. For most systems the sticking probability varies with coverage and temperature. Frequently the form of plots of s against θ is that they start with an initial value, s_0, the initial sticking probability; as the coverage is increased s gradually decreases towards zero as the adsorbate–surface and adsorbate–adsorbate interactions change. In studies of the temperature dependence of s_0, there are two common outcomes; either it does not vary very much with temperature for a given adsorption system or it is found to decrease with increase in temperature.

3.2.1 The precursor state

The characteristic shape of sticking probability curves is generally determined by the 'accommodation' of a molecule in a weakly bound physisorbed precursor state prior to chemisorption. The precursor model was first postulated by Kisliuk in 1957. He showed that the sticking coefficient could be related to the probabilities of the chemisorption, desorption and migration of a molecule from above empty and filled chemisorption sites. Taking into account the fractional coverage adsorbed on the surface, he obtained an expression for s for chemisorption where the gas molecule adsorbs on a single site on the surface:

$$s = \frac{p_a(1 - \theta)}{1 - p_m} \tag{3.15}$$

where p_a is the probability of chemisorption above an empty site and p_m is the probability of migration of the molecule across the surface from a starting position above an empty site (θ is the fractional coverage, see eqn (3.2)). When a molecule migrates to a second site on the surface, it can once again chemisorb, desorb or migrate, just as it could from the first site.

The sum of the probabilities of these processes occurring from above an empty site must equal one, that is

$$p_a + p_d + p_m = 1 \tag{3.16}$$

where p_d is the probability of desorption from above an empty site. The corresponding probabilities of chemisorption, p_a', desorption, p_d', and migration, p_m', for a molecule in a filled adsorption site are

$$p_a' = 0 \quad \text{and} \quad p_d' + p_m' = 1 \tag{3.17}$$

because a molecule cannot chemisorb on a site which is already filled.

If we now consider chemisorption on a surface where the fractional coverage is θ: the probability that a molecule is to be found above an occupied chemisorption site is θ, while the probability that it is above an unoccupied site is $(1 - \theta)$. The probability of adsorption on the first site visited, $p_{a(1st\ site)}$, is therefore $p_{a(1st\ site)} = p_a(1 - \theta)$. The probability of desorption from the first site visited, $p_{d(1st\ site)}$, is the sum of the probabilities of desorption from above empty and filled sites, which is given by $p_{d(1st\ site)} = p_d(1 - \theta) + p_d'$. Substitution of these values for p_a and p_d in eqn (3.16) gives the probability of migration from the first unfilled site visited:

$$p_{m(1st\ site)} = 1 - p_a - p_d + \theta(p_a + p_d - p_d') \tag{3.18}$$

At zero coverage, the initial sticking probability, s_0, is obtained from substituting $\theta = 0$ into eqn (3.15),

$$s_0 = \frac{p_a}{1 - p_m} = \frac{p_a}{p_a + p_d} \tag{3.19}$$

Taking the ratio of eqn (3.15) to eqn (3.19) then gives

$$\frac{s}{s_0} = \left(1 + \frac{K\theta}{1 - \theta}\right)^{-1} \tag{3.20}$$

where

$$K = \frac{p_d'}{p_a + p_d} \tag{3.21}$$

Figure 3.10(a) shows the expected forms of the sticking probability curves for different values of K. The data for 'real' adsorption systems usually correspond to curves where $K < 1$.

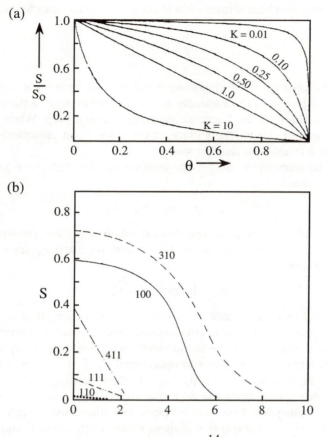

Fig. 3.10 (a) The Kisliuk curves which model the sticking probability curves for one site adsorption processes. (From Kisliuk 1957.) (b) The sticking probability as a function of coverage for N_2 adsorption on a number of W faces. (From Singh-Boparai and King 1975.)

For some systems, the molecule needs two adjacent sites in order to adsorb on the surface. In this case the s/s_0 ratio is given by

$$\frac{s}{s_0} = \frac{(1 - \theta)^2}{1 - \theta(1 - K^*) + \theta^2 s_0}$$

where

$$K^* = \frac{p_{a2} - p_{a1}}{p_{a1} - p_{a2}} \tag{3.22}$$

The subscripts a1 and a2 refer to the two binding sites. Generally, $1 > K^* > -s_0 > -1$.

3.2.2 Temperature dependence

Both the initial sticking probability and the shape of the sticking probability versus coverage curve may be temperature dependent. This is thought to be due to the pre-equilibrium between the gas phase and the precursor state; the system is usually analysed by combining the probability approach adopted in Section 3.2.1 with the rate constant. Assuming that the probabilities of surface processes are proportional to their rates, then eqn (3.19) can be rewritten

$$s_0 = \frac{k_a}{k_a + k_d} \tag{3.23}$$

where k_a and k_d are the rate constants for chemisorption and desorption of the precursor, respectively. This can be rearranged to give

$$\frac{k_d}{k_a} = \frac{1 - s_0}{s_0} \tag{3.24}$$

Each of the rate constants can be described in terms of the usual type of exponential expressions:

$$k_d = A_d \, \exp\left(\frac{-E_d}{RT}\right) \text{ and } K_a = A_a \, \exp\left(\frac{-E_a}{RT}\right) \tag{3.25}$$

where A denotes the pre-exponential factor and E the activation energies for desorption (subscript d) and adsorption (subscript a). R and T have their usual meanings. These expressions can be substituted into eqn (3.24) and when natural logs are taken, the following relationship is obtained, relating the activation energy barriers to the initial sticking coefficients.

$$\ln\left(\frac{A_d}{A_a}\right) + \left(\frac{E_a - E_d}{RT}\right) = \left(\frac{1 - s_0}{s_0}\right) \tag{3.26}$$

If we make the reasonable assumption that the pre-exponential factors are rather temperature independent, we can see that the temperature dependence of the initial sticking probability is governed by the difference in activation energies of adsorption and desorption.

(i) For non-activated chemisorption $E_a < E_d$. This means that s_0 increases, and thus $(1 - s_0)/s_0$ decreases, as the temperature decreases.

(ii) For activated chemisorption $E_a > E_d$, which means that s_0 decreases as the temperature increases. This situation is observed far less frequently.

Generally, the sticking probability is often found to be different for different faces of the same metal crystal; this might be because of the nature of the adsorption and precursor site(s) available on the substrate. Repulsive/attractive interactions between adjacent adsorbate atoms/molecules are also found to be of importance in some cases.

The specificity of adsorption on a certain crystal surface is clearly demonstrated by the sticking probabilities for dissociative chemisorption of molecular nitrogen on different faces of tungsten, which are shown as a function of surface coverage in Fig. 3.10(b). There are two types of curve in this case. For some faces, a linear dependence of sticking probability with coverage is observed, e.g. nitrogen on W(411) and (111). For others such as nitrogen on W(310) and (100), the initial sticking probability is reasonably independent of nitrogen coverage in the low-coverage region, and then decreases with the sigmoid shape shown. The Kisliuk type of model for two adjacent sites can be used to model this latter behaviour, modified to take into account a pairwise repulsion effect between adjacent adsorbed nitrogen atoms. The structure of the adsorbed layer leads to small additional modifications to the Kisliuk model.

3.3 Adsorption isotherms and rates

An adsorption isotherm is a plot of the equilibrium amount of a substance adsorbed (in terms of coverage or uptake) against pressure of the adsorbing substance in the gas phase, measured at constant temperature. Most adsorption systems fall into one of the five categories which are sketched in Fig. 3.11.

Measurements of these isotherms can be made using either gravimetric or volumetric techniques and it turns out that types I and II are of most interest in surface science studies.

In the type I case, the amount adsorbed increases steadily with pressure until a plateau is reached at $\theta = 1$. No further adsorption occurs at this stage. This isotherm describes 'ideal' chemisorption, where molecules chemisorb until the surface becomes saturated with adsorbate, whereupon adsorption ceases.

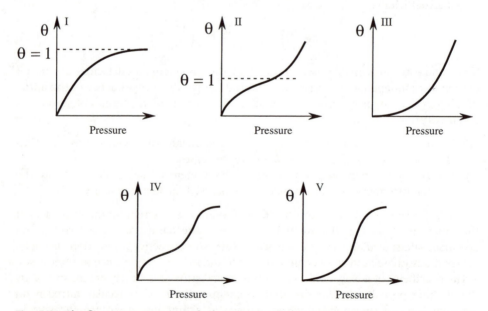

Fig. 3.11 The five main categories of adsorption isotherms, given as plots of coverage, θ, against equilibrium pressure.

In the type II case, as the monolayer plateau region is approached, there is a further increase in the amount adsorbed and many layers are ultimately adsorbed. This type of isotherm is usually associated with physisorption.

Of the other three categories, type III is associated with multilayer formation from the onset, and types IV and V are associated with the formation of monolayers where there are variations of heat of adsorption with θ. Category IV behaviour is of considerable interest in the study of real catalysts as it implies 'hysteresis behaviour' which means that an increase in pressure to a certain level results in a different coverage when compared to the coverage formed by decreasing the pressure to the same level. This behaviour is usually observed for evaporation and condensation in pores such as those commonly found in catalysts and arises because evaporation in a porous material is not as easy as condensation, as there is a high probability of recondensation. It follows that the shape of the hysteresis can often indicate the size and type of pore in the catalyst itself.

To return to the type I and II isotherms, which apply more appropriately to studies of the flat surface, models to describe the observed behaviour have been set up. The first, the Langmuir adsorption isotherm, describes type I behaviour, whereas the Brunauer–Emmett–Teller isotherm (BET) extends the Langmuir model to reproduce physisorption phenomena of type II.

3.3.1 The Langmuir isotherm

This is the most commonly used model for an adsorption isotherm and describes ideal chemisorption systems. It can be derived on the basis of kinetic, thermodynamic, and/or statistical mechanical models. It will be derived here in kinetic terms. The assumptions made in this model are that:

(i) adsorption occurs on specific sites and all adsorption sites are identical;
(ii) the energy for adsorption is independent of how many of the surrounding sites are occupied;
(iii) only one adsorbate occupies each site and once all the sites are occupied, adsorption ceases (i.e. a maximum of one monolayer is deposited).

It is important to realize that very few systems give rise to Langmuir-like behaviour. The first and third assumptions, which imply that only one type of site is occupied, are often wrong. LEED measurements show that there are several possible binding sites available, even on the simplest, low-indexed faces. Techniques which are used to observe adsorbates, such as those discussed later in this chapter, have often revealed that more than one of these sites can be occupied in a given adsorption system. In addition, even single-crystal surfaces contain defects/step edges and these tend to be occupied first.

The second assumption is nearly always wrong. As we have seen in Sections 3.1.4 and 3.2, the heat of adsorption and the sticking probability are almost always coverage dependent. The main influences on the heat of adsorption are the total amount of charge transfer between the adsorbate and the surface, which obviously changes with coverage, and the lateral interactions between adsorbates, which are particularly noticeable at higher coverages.

Another significant drawback of the Langmuir isotherm is that it does not take into account ordering of the adsorbate layer as it is deposited on the surface, or of adsorbate-induced reconstruction of the surface.

Despite the limitations of the Langmuir isotherm, it continues to be widely used. This is because it is a simple model which can help to deduce quantitative relationships between the amount of adsorbate on the surface and the pressure in the gas phase above it. In principle, the Langmuir isotherm contains all of the parameters which are needed to do this, and it provides a good first approximation. The Langmuir isotherm essentially provides a framework for describing the extent and strength of adsorption on a surface, it sets a reasonable basis for a useful method of determining surface kinetics and it can also be used to determine the surface areas of solids (it is particularly useful in this respect for studying real catalysts).

There are several modified isotherms that attempt to take into account the limitations of the Langmuir isotherm and three of these are discussed later in this section.

3.3.1.1 *The Langmuir isotherm for a single adsorbing species*

The Langmuir isotherm is derived kinetically, assuming that when the system comes to equilibrium, the rates of adsorption and desorption are equal and finite. In order to do this the rate expressions for adsorption and desorption of the adsorbate are equated. We will consider the situation for adsorbing molecules of a gas A onto the surface such that

$$A_{(gas)} + Surface \rightarrow A_{(ads)} \tag{3.27}$$

Figure 3.12 shows a typical potential energy curve involved. This figure shows that the activation energy barrier for chemisorption, E_a, is typically quite small. Once in the

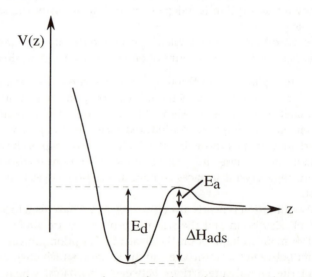

Fig. 3.12 The potential energy curve showing the relationship between the activation barriers for desorption, E_d, adsorption, E_a and the heat of adsorption, ΔH_{ads} (or q).

chemisorption well, the activation energy for desorption from the well, E_d, is much larger. The difference between the two constitutes the heat of adsorption ΔH_{ads}, that is

$$\Delta H_{ads} = E_d - E_a \tag{3.28}$$

First an expression must be found for the rate of adsorption per unit area, R_{ads}. This is given by

$$R_{ads} = \left[\frac{P}{(2\pi mkT)^{\frac{1}{2}}} \right] (1 - \theta) \exp\left(\frac{-E_a}{RT} \right) \tag{3.29}$$

The first part of the expression is the collision rate (in terms of collisions per second per unit area) of the adsorbate gas with the surface, $dN/dt = P/(2\pi mkT)^{\frac{1}{2}}$, which was given in Section 3.1.5. (eqn (3.13)).

The second part of the expression in eqn (3.29), $(1 - \theta)$, describes the probability of hitting a vacant site so that adsorption can take place. The final part of the expression is the Boltzmann factor, which takes into account the activation energy, E_a.

A similar expression can be deduced for desorption:

$$R_{des} = \theta N_s A \, \exp\left(\frac{-E_a}{RT} \right) \tag{3.30}$$

which assumes first-order desorption kinetics. N_s is the concentration of surface sites per unit area and A is a pre-exponential or frequency factor.

If the adsorption and desorption expressions are now equated and rearranged to give an expression for the coverage:

$$\theta = \frac{P(1 - \theta)}{(2\pi mkT)^{\frac{1}{2}} N_s} A \, \exp\left(\frac{-E_a}{RT} + \frac{E_d}{RT} \right) \tag{3.31}$$

We define b, the Langmuir b factor, by

$$b = \frac{\exp\left(\frac{E_d - E_a}{RT} \right)}{(2\pi mkT)^{\frac{1}{2}} N_s A} = \text{const} \times \exp(q/RT) \tag{3.32}$$

where const is the product of all the constants. The expression in eqn (3.31) then simplifies to

$$bP = \frac{\theta}{1 - \theta} \quad \text{or} \quad \theta = \frac{bP}{1 + bP} \tag{3.33}$$

which are alternative statements of the Langmuir isotherm. We can consider two extreme cases.

(i) At low pressure, $bP << 1$, from the expression (3.33), $\theta \sim bP$, so the coverage is proportional to the pressure, as we see for the type I curve in Fig. 3.8.

(ii) At high pressure, $bP >> 1$, eqn (3.33) reduces to $\theta \sim 1$, which models the asymptotic region of the type I curve in Fig. 3.11.

The Langmuir isotherm therefore successfully models type I behaviour for chemisorption up to one monolayer.

When results are obtained from volumetric measurements, θ is usually found from the ratio of the volume adsorbed, V, to the volume needed to produce a coverage of one monolayer, V_m. In this case, using the Langmuir isotherm model, the substitution $\theta = V/V_m$ is used so that

$$V/V_m = \frac{bP}{1 + bP}$$
$$\Rightarrow bP/V = \frac{1 + bP}{V_m} \qquad (3.34)$$
$$\Rightarrow P/V = \frac{P}{V_m} + \frac{1}{bV_m}$$

If the system conforms to the Langmuir isotherm, a plot of P/V against P would give a straight line of slope $1/V_m$ and intercept $1/bV_m$.

3.3.1.2 The Langmuir isotherm for two co-adsorbing species

The Langmuir isotherm can also be used to describe the adsorption of two gases, A and B, which are competing to adsorb on the surface.

The outcome of the adsorption process in this case depends on the pressures of the two gases and the relative heats of adsorption; the heats of adsorption are accounted for in the Langmuir b factors of the two gases, b_A and b_B.

The expressions for the coverages of A and B are

$$\theta_A = \frac{b_A P_A}{1 + b_A P_A + b_B P_B} \qquad \theta_B = \frac{b_B P_B}{1 + b_A P_A + b_B P_B} \qquad (3.35)$$

When the pressures of the two components are equal, the ratio of the coverages are given by the ratio of the b factors.

If $b_A >> b_B$ the expression for θ_B becomes

$$\theta_B = \frac{b_B P_B}{b_A P_A} \qquad (3.36)$$

3.3.1.3 The Langmuir isotherm for a molecule which dissociates on adsorption

When the following process takes place, the Langmuir isotherm can still be applied:

$$AB_{(gas)} + \text{Surface} \rightarrow A_{(ads)} + B_{(ads)} \qquad (3.37)$$

In this case the expression has to be modified because for a single molecule dissociating, two sites must be available on the surface. The probability of finding one vacant site on the surface is $(1 - \theta)$ and so the probability of finding two is $(1 - \theta)^2$. The situation for desorption is also slightly different in that the probability of two particles coming together to form the desorbing molecule must be included. This depends on the concentration of each and so is proportional to θ^2. By making these modifications in the initial expressions for the rates of adsorption and desorption, the following expression is obtained for the dissociating molecule:

$$P = \frac{1}{b}\left(\frac{\theta}{1-\theta}\right)^2 \quad \text{or} \quad \theta = \frac{(bP)^{\frac{1}{2}}}{1+(bP)^{\frac{1}{2}}}. \tag{3.38}$$

3.3.2 The Brunauer–Emmett–Teller isotherm (BET)

The Langmuir isotherm is limited to the adsorption of one monolayer of adsorbate. In a system where the latent heat of vaporization of the adsorbate is significantly smaller than the heat of adsorption, adsorption can take place at temperatures well above the boiling point of the adsorbate, and the formation of further layers is very unlikely. If the heat of adsorption and the latent heat of vaporization are similar, then a low temperature is needed to produce significant adsorption in the first layer and these conditions may result in the formation of further layers above it. The formation of multilayers is modelled by the Brunauer–Emmett–Teller (BET) isotherm and this essentially uses the basis of the Langmuir model and extends it to encompass physisorbed multilayers. It therefore models type II behaviour. The assumptions made are that:

(i) each layer of adsorbate is treated as a Langmuir monolayer and each layer must be complete before the next layer starts to form;
(ii) the heat of adsorption for the first layer, ΔH_{ads1}, is characteristic of the adsorbate/adsorbent system;
(iii) the heat of adsorption for subsequent layers is equal to the heat of condensation, ΔH_L, such that $\Delta H_L = \Delta H_2 = \Delta H_3 = \Delta H_4 = \Delta H_5 = \Delta H_6 = \ldots$.

The general form of the BET isotherm is

$$\theta = \frac{c(P/P_0)}{\left(1 - \frac{P}{P_0}\right)(1 + (c-1)P/P_0)} \tag{3.39}$$

P is the equilibrium pressure over the solid and P_0 the saturated vapour pressure of the gas at the temperature of the experiment. The term c is defined as

$$c = \exp\left(\frac{\Delta H_1 - \Delta H_n}{RT}\right) \tag{3.40}$$

where ΔH_n is the heat of adsorption in subsequent layers. When $\Delta H_1 \gg \Delta H_n$ the BET isotherm reduces to the Langmuir isotherm.

If volumetric measurements ($\theta = V/V_m$) are taken then the expression given in eqn (3.39) can be rewritten

$$\frac{P}{V(P_0 - P)} = \frac{1}{V_m c} + \frac{(c-1)P}{V_m c P_0} \tag{3.41}$$

so a plot of $P/V(P_0 - P)$ against P/P_0 is a straight line if the system conforms to the BET isotherm. The slope $= (c-1)/(cV_m)$ and the intercept is at $1/(V_m c)$.

The value of c gives insights into the nature of the adsorption system. Large values of c show a large difference of heat of adsorption between first and subsequent layers, as shown in Fig. 3.13 for the case of $c \approx 1000$ and $c \approx 200$. For large values of c, it is

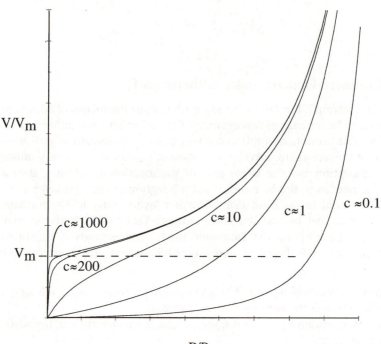

Fig. 3.13 Plots of V/V_m against P/P_0 showing the effect of different values of c on the determination of V_m. V_m is indicated on the figure.

relatively easy to find a value for V_m. As the difference between ΔH_{ads} for the first and subsequent layers decreases, such that $c \approx 10$ to $c \approx 0.1$, the asymptotic region is far less well defined and V_m is much more difficult to distinguish. It should be noted that physisorption in the first layer is often slightly stronger than that in subsequent layers.

The BET isotherm is important because it is widely used to measure surface areas, in particular of thin films and powders. The value of V_m is obtained by measuring the uptake of an inert gas or molecular nitrogen at low temperature. An assumption is then made about the packing of the adsorbates on the surface and the area each occupies, and the surface area is then determined from the data.

3.3.3 Other isotherms

There are several other isotherms, largely based on the Langmuir isotherm, which have been developed to take into account the limitations of the Langmuir model. Two examples are described briefly here.

3.3.3.1 *The Freundlich isotherm*

This form of the isotherm was developed to take into account the observation of a coverage dependence of the heat of adsorption. Empirically it is given by

$$\theta = kP^{\frac{1}{n}} \tag{3.42}$$

where k and n are constants and $n > 1$. The assumptions made in the Freundlich isotherm are that:

(i) the heat of adsorption declines logarithmically with coverage so $\Delta H = -\Delta H_m \ln \theta$ (ΔH_m, the heat of adsorption for the monolayer, is a constant for a given system);

(ii) θ has values which do not approach 0 or 1, that is it applies for intermediate coverages only, usually in the range 0.2–0.8.

This isotherm can be applied to both chemisorption and physisorption systems.

3.3.3.2 *The Temkin isotherm*

This isotherm applies only to monolayer chemisorption and was devised to take into account the fact that not all sites are energetically equivalent. The heat of adsorption is assumed to decrease linearly with coverage such that

$$\Delta H_{ads} = \Delta H_0 (1 - \beta\theta) \tag{3.43}$$

where β is a constant for a given system. The Temkin isotherm is written empirically as

$$\theta = k_1 \ln (k_2 P) \tag{3.44}$$

where k_1 can be shown to be related to the heat of adsorption by the equation $k_1 = -RT/\beta \, \Delta H_0$, and k_2 is a constant, also related to the heat of adsorption.

3.4 Measuring heats of adsorption, isosteres and desorption rates

The heat of adsorption was described and defined early in this chapter (in Sections 3.1.2 and 3.3.1.1). It is useful now to look how it can be measured and related to the isotherms of the previous section.

3.4.1 The isosteric heat of adsorption

The equilibrium between the vapour phase and the adsorbed phase can be described in terms of simple classical thermodynamics, using the Clausius–Clapeyron equation

$$\frac{\mathrm{d} \ln P}{\mathrm{d}\left(\frac{1}{T}\right)} = \frac{\Delta H_{ads,\theta}}{R} = -\frac{q}{R} \tag{3.45}$$

where P is the vapour pressure. When applied to a surface–adsorbate system, the coverage at which the heat of adsorption is measured must be defined because of its coverage dependence. ΔH_{ads} can be determined if P can be measured at a number of

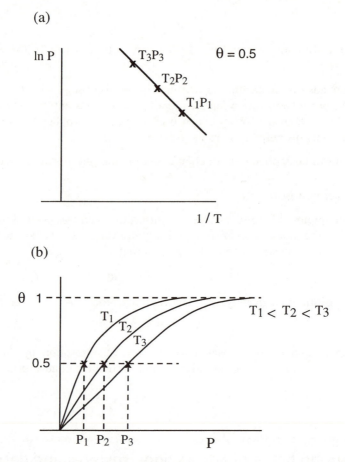

Fig. 3.14 (a) A plot to determine the isosteric heat of adsorption at a given coverage (here, $\theta = 0.5$) and (b) the related isotherms, where $T_1 < T_2 < T_3$.

different temperatures. A plot of ln P against $1/T$, the isostere, is given in Fig. 3.14(a), together with the related isotherms in Fig. 3.14(b). The data needed for plotting the isostere can be taken for points of equal coverage such as 0.5 as the figure shows. In this case the different temperatures $T_1 < T_2 < T_3$ are indicated on the isotherms with the associated pressure measurements P_1, P_2 and P_3. The points resulting from them are given on the isostere for $\theta = 0.5$. The slope of the isostere $= \Delta H_{\text{ads},\theta}/R$. If ΔH_{ads} is found in this way, it is called the *isosteric heat of adsorption*.

3.4.2 Temperature programmed desorption (TPD)

Another method of measuring the approximate heat of adsorption uses kinetic information obtained by rapidly heating the surface to above the temperature at which the adsorbed layer desorbs. The pressure burst of desorbing gas as a function of substrate temperature is recorded by a mass spectrometer in this technique, which is called temperature programmed desorption (TPD) or thermal desorption spectroscopy (TDS). Typical heating rates give a substrate temperature rise of 10–30 K s^{-1}. The general rate expression for this process can be written as

$$-\frac{dN_{ads}}{dt} = A N_{ads}^n \exp\left(\frac{-E_d}{RT}\right) \tag{3.46}$$

where A is the pre-exponential (or frequency) factor, N_{ads} the concentration of adsorbate molecules on the surface and E_d the activation energy for desorption. n is the order of the desorption process. From the desorption curves recorded for different initial coverages of adsorbate, the values of A and n can be deduced and thus E_d can be calculated.

It should be noted that in the situation where the activation energy barrier for adsorption is zero, from Fig. 3.12 and eqn (3.28), it can be seen that $E_d = -\Delta H_{ads} = \Delta H_{des}$ and the heat of adsorption can be found directly from the TPD data.

The linear heating rate dT/dt is related to the rate of desorption by

$$\frac{-dN_{ads}}{dt} = -\frac{dT}{dt}\ \frac{dN_{ads}}{dT} \tag{3.47}$$

Equation (3.46) can therefore be written in terms of dN_{ads}/dT as

$$\frac{-dN_{ads}}{dT} = \frac{dt}{dT}\ A N_{ads}^n \exp\left(\frac{-E_d}{RT}\right) \tag{3.48}$$

The maximum rate of desorption can be found at the point at which both d^2N_{ads}/dt^2 and d^2N_{ads}/dT^2 are zero. Differentiation of eqn (3.48) and setting it to zero gives the value of T_{max}, the temperature at which the peak of desorption is observed in the TPD spectrum:

$$n N_{ads}^{n-1} \frac{dN_{ads}}{dT} = -\frac{N_{ads}^n E_d}{RT_{max}^2} \tag{3.49}$$

The substitution for dN_{ads}/dT from eqn (3.48) into eqn (3.49), results in

$$\frac{A n N_{ads}^{n-1} dt}{dT} \exp\left(\frac{-E_d}{RT_{max}}\right) = \frac{E_d}{RT_{max}^2} \tag{3.50}$$

This expression can be used to show the characteristics of the TPD curves which are observed for different orders of desorption process. Equation (3.50) shows that, generally, T_{max} decreases with increasing order of reaction.

(i) First-order desorption occurs for $n = 1$. Equation (3.50) then reduces to

$$\frac{A\, dt}{dT} \exp\left(\frac{-E_d}{RT_{max}}\right) = \frac{E_d}{RT_{max}^2} \tag{3.51}$$

which means that T_{max} is independent of the initial coverage. The expected form of the TPD curves for a first-order desorption process are illustrated in Fig. 3.15(a), showing that the peak of desorption occurs at the same temperature for each initial coverage. The shapes of the peaks are also closely similar to each other and

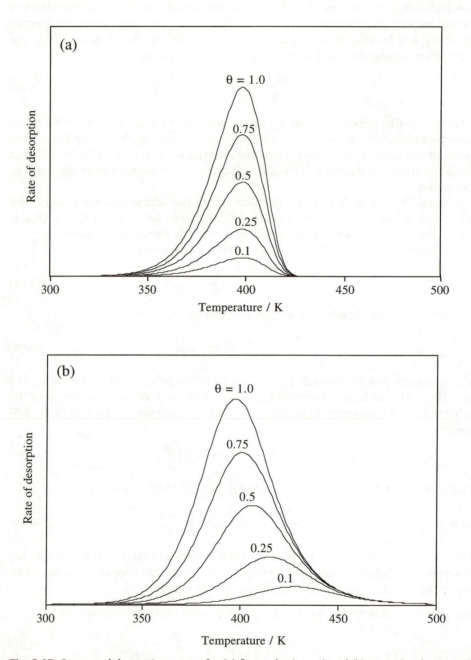

Fig. 3.15 Computed desorption curves for (a) first order ($n = 1$) and (b) second order ($n = 2$) desorption from a surface. The plots were calculated using identical linear heating rates and values of E_d and are plotted as rate of desorption as a function of temperature for a range of coverages.

are asymmetric; the curve has a steady slope up to the peak and then falls away rapidly at temperatures above the peak.

(ii) Second-order desorption occurs when $n = 2$. For this, eqn (3.50) becomes

$$A\,2\,N_{T\,max}\,\frac{dt}{dT}\,\exp\!\left(\frac{-E_d}{RT_{max}}\right) = \frac{E_d}{RT^2_{max}} \tag{3.52}$$

which shows that the coverage $N_{T\,max}$, the coverage at T_{max}, is related to T_{max}. As the initial coverage is increased, T_{max} decreases. The second-order TPD curves are very symmetrical as illustrated in Fig. 3.15(b).

Looking at how the peak temperature changes as a function of initial coverage should therefore indicate the order of the desorption process. However, the situation can be complicated if the activation energy for desorption changes with coverage (which is quite common). If it decreases with coverage for a first-order desorption process, then the desorption maximum will shift down in temperature with increasing coverage, and this could be mistaken for a second-order process. It is therefore important to bear in mind the shapes of the desorption curves as well as the behaviour of T_{max}.

From eqns (3.51) and (3.52), estimates of E_d can be made by assuming a value of A. For first-order reactions, A is usually taken as being in the range 10^{13}–10^{15} s^{-1} (that is of the order of magnitude of the adsorbate to surface vibration frequency) and is obtained from considerations relating to the pre-exponential factors in gas-phase reactions. Assuming a value of A is not altogether satisfactory and so the assumed value is usually quoted.

A better way of finding E_d from the data is to use information from different heating rates.

(i) For first-order desorption, taking logs of eqn (3.51) and differentiating with respect to T_{max} gives

$$\frac{d(\ln\,dT/dt)}{d(\ln T_{max})} = 2 + \frac{E_d}{RT_{max}} \tag{3.53}$$

so a plot of $\ln\,dT/dt$ against $\ln T_{max}$ is linear for first-order desorption and E_d can be found from the gradient.

(ii) For the second-order desorption process, similar treatment of eqn (3.52) gives

$$\ln(N_{init}\,T^2_{max}) = \frac{E_d}{RT_{max}} - \ln\frac{dt\,AR}{dT\,E_d} \tag{3.54}$$

where N_{init} is the initial coverage of the surface (it is known that $N_{T\,max} = N_{init}/2$). A plot of $\ln(N_{init}T^2_{max})$ against $1/T_{max}$ is linear for second-order kinetics and again the slope can be used to obtain the value of E_d. In this case the intercept can be used to determine A.

The TPD data become more complicated if the adsorbate is adsorbed in more than one state. Obviously the most weakly bound state will be desorbed first, and the different states may desorb with different orders. It should also be noted that the same factors which may cause a strong coverage dependence for ΔH_{ads} also effect E_d, namely total charge transfer between the adsorbate and surface, and lateral interactions of the adsorbates themselves.

3.4.3 Microcalorimetry

Both the Clausius–Clapeyron analysis of equilibrium adsorption isosteres and TPD measurements provide indirect means of determining the heats of adsorption. The main limitation of both of these methods is that the adsorbate species being studied must be reversibly adsorbed on the surface. This means that values of ΔH_{ads} cannot be obtained for the most catalytically interesting systems, where a surface reaction takes place and a 'new' species, the product, is desorbed. Moreover, the methods cannot be applied to many oxide systems, which provide some of the most useful models for real catalysts and corrosion systems, because oxides are often found to 'dissolve' into the bulk when heated.

A further problem with TPD is that in order to accurately determine ΔH_{ads}, there must be a precise knowledge of the desorption rate law (which may well be complex) and the variation of the kinetic parameters as a function of the coverage.

Calorimetry provides a direct method for determining the heat of adsorption and also the sticking probability, but it is extraordinarily difficult to perform experimentally on single-crystal samples because of the very small sizes of the samples. The demanding nature of the experiment can be simply illustrated by the consideration that on a single-crystal surface of only a few square millimetres, the adsorption of submonolayers of gas typically releases well below 1 μJ of energy. However, a technique has been developed and a some very valuable studies have been carried out as a result of the pioneering work of King in the early 1990s.

In calorimetric measurements, the output of energy from the process being studied is monitored by observing the change in temperature of either the reacting sample or the reaction vessel containing the reaction system. The experiments are designed so that the heat capacity of the sample/apparatus is known and this, in combination with the observed temperature rise, can be converted to give a value for the enthalpy. The small energy changes associated with adsorption on the surface mean that in order to measure a finite temperature change, the sample must be of very low thermal mass compared to its surface area, and a rapid adsorption process must take place. In King's experiments, an unsupported single crystal is grown as a thin film only about 200 nm thick. This is usually done by growing the film on a crystal face of NaCl, which acts as a template for the metal. The NaCl is then dissolved away and the crystal is cold-welded onto a hollow ring of Ni which acts as a support. Once in vacuum, a supersonic beam of adsorbate gas is pulsed onto the surface and the temperature rise is recorded using a very sensitive mercury cadmium telluride (MCT) photoconductive detector.

An example of the data obtained for the adsorption of oxygen on Ni(100) is shown in Fig. 3.16. Both the heats of adsorption and the sticking probabilities have been extracted from the data as a function of adsorbate coverage. This provides an excellent means of correlating adsorption kinetics with energetics. In the case of oxygen on Ni, ΔH_{ads} decreases very slowly below 0.1 ML coverage and then drops rapidly between $\theta = 0.1$ and $\theta = 0.15$. From $\theta = 0.15$ to 0.35, it is again found to decrease slowly. The sticking probability decreases rapidly at low coverage and then more steadily towards zero at $\theta = 0.35$.

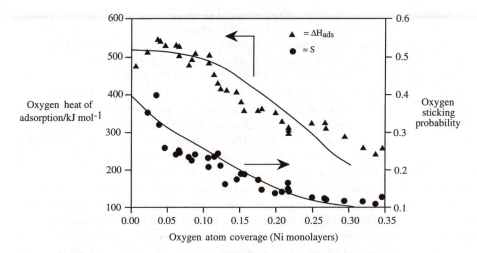

Fig. 3.16 Heat of adsorption (filled triangles) and sticking probability (filled circles) measured using a microcalorimeter, for oxygen chemisorption on Ni(100). Simulated results (solid line) are also shown. (From Al-Sarraf *et al.* 1993.)

3.5 Adsorption sites and geometries

An important aspect of the adsorption process is the nature of the site in which the adsorbate is bound. The sites available have considerable influence on the binding of an adsorbate and on the charge transfer that takes place between the adsorbate and the surface. Indeed, different faces of the same crystal might exhibit different behaviour with respect to certain adsorbates because not only might one be more open in structure than another (with associated differences in electron density in the valence band) but also, in some cases, one might offer less favourable sites for binding than the other. For example, an fcc (100) face does not offer any threefold bridging sites, whereas a (111) surface does not offer any fourfold ones.

In this section we will look first at the vibrational spectroscopies, which have provided valuable insights into the sites and binding geometries of adsorbed species, before looking at the binding of CO, the favourite adsorbed species of the surface scientist. Hydrocarbon adsorption on surfaces will also be described as this encompasses a wide range of possible binding configurations and is important in industrial processes. Some adsorbates are found to induce reconstructions in the surface, and this will also be considered. In all cases it should be noted that surface scientists use and require a number of complementary techniques to properly study an adsorption system.

3.5.1 Vibrational spectroscopies

The attraction of using vibrational spectroscopies to study adsorbates on surfaces lies in the fact that there is, for comparison using group frequencies, a huge database of IR and Raman spectra of numerous compounds in the solid, liquid and gaseous phases. Vibrational data on the adsorbates can be used to determine the bonding pattern of the adsorbed species, including information about any fragmentation which has taken place. In addition, they can often be used to determine what sort of site the adsorbate has adopted.

Two main techniques have been developed to study vibrations of monolayers of species adsorbed on single-crystal surfaces and these are high-resolution or vibrational electron energy loss spectroscopy (HREELS or VEELS) and reflection–absorption infrared spectroscopy (RAIRS). A serious difficulty encountered experimentally is that of sensitivity, because sample sizes are so small. A typical single-crystal sample has an area of the order of 1 cm^2. In Section 2.1.3, we calculated N_s to be of the order of 10^{15} atoms cm^{-2} which means that in real terms the sample contains about 10^{15} adsorbate atoms or molecules. This is in comparison to solid-phase samples studied by vibrational spectroscopy, which typically contain about 10^{19} molecules. VEELS was the first technique to be developed for studying adsorbates as higher sensitivity could be obtained by using electrons as the probe species. However, advances in IR instrumentation using Fourier transform methods have enabled RAIRS to provide adequate sensitivity with much higher resolution.

3.5.1.1 Vibrational electron energy loss spectroscopy (VEELS)

In this technique a beam of 'monoenergetic' electrons with energy E_0 (typically in the range of about 1–10 eV) is fired at a surface, as shown in Fig. 3.17(a).

The zero-order diffraction beam that is scattered off the surface in the roughly specular direction is collected as shown. It is found that a very small proportion of the collected electrons have lost (or to a lesser extent gained) energy, ΔE. The energy lost is due to the absorption of quanta of vibrational energy by the adsorbate or the surface such that

$$\Delta E = h\nu_{\text{vibrational}} \tag{3.55}$$

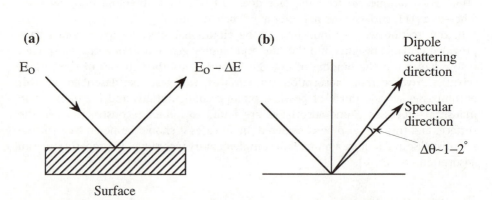

Fig. 3.17 The principle of EELS. (a) When a 'monoenergetic' beam of electrons is incident on a surface, most of the beam is elastically scattered; however, a small proportion of the electrons lose (or gain) energy ΔE. (b) These inelastically scattered electrons are found to be scattered slightly away from the specular direction as shown.

where h is the Planck constant and $\nu_{\text{vibrational}}$ is the vibrational frequency. In electron spectroscopies such as XPES, typical resolutions required are about 0.5–1 eV to adequately resolve the peaks. For VEELS the required resolution is much higher, at about 1–10 meV (equivalent to 8–80 cm^{-1} in wavenumber units of cm^{-1}; these are conventionally used by vibrational spectroscopists), because of the relative magnitudes of electronic and vibrational transitions. Vibrational transitions occur typically in the wavenumber range 4000–100 cm^{-1} (1 meV = 8.067 cm^{-1}).

It is usual to record only the 'loss' side of the VEELS spectrum because there is a higher probability of the electrons losing energy than gaining it (by analogy with Raman spectroscopy where Stokes scattering is more intense than Antistokes). A typical VEELS analyser is shown in Fig. 3.18. There are a number of designs of the VEELS analyser but the principle of their operation is the same. The electron source produces electrons which have a Boltzmann distribution of energies. Prior to hitting the surface, the beam must be 'monochromated' as much as possible, as the range of energies within the incident beam largely determines the resolution. The monochromator consists of a series of sectors which are set at different potentials to 'select out' a narrow range of energies (usually in the range equivalent to 20–80 cm^{-1}, although the most advanced spectrometers can now obtain better than 8 cm^{-1}). The monoenergetic beam hits the surface and is scattered into the analyser which consists of a set of sectors similar to that in the monochromator. The potentials applied to the sectors in the analyser allow the energies of the reflected electrons to be scanned over the appropriate energy range and the numbers of these are measured by the electron multiplier. The spectrum is recorded in the form $N(E)$ against E, as shown schematically in Fig. 3.19. Two vibrational quanta, $\Delta\nu_1$ and $\Delta\nu_2$, are indicated on the figure.

The first VEEL spectra recorded were much simpler than had been expected. It turned out that this was due to the dominant interaction mechanism of the electrons with the surface–adsorbate systems. This most prominent scattering mechanism is 'dipole scattering' and for metal surfaces, this obeys the **metal surface selection rule** (termed MSSR), which states that: *Only those totally symmetrical vibrational modes which have an oscillating dipole perpendicular to the metal surface will be allowed (or excited).*

The dipole scattering mechanism is due to a long-range interaction involving the electric fields from the incoming electrons and the adsorbate dipoles, which vary in magnitude during vibration. Because the conduction-band electrons are free to move, if a charged species sits above the surface, as shown in Fig. 3.20(a), it produces a self-image of equal size and opposite sign, within the surface, in a so-called '*image-charge*' effect. We can now consider the response of the conduction-band electrons to the presence of a dipole which is perpendicular to (Fig. 3.20(b)) and parallel to (Fig. 3.20(c)) the surface. In the perpendicular case, the image charge effect gives a dipole which is enhanced by a factor of two, so that when it acts as a dynamic dipole (i.e. when it oscillates) it interacts strongly with the electromagnetic field of the incoming electrons. When the dipole is parallel to the surface as shown in Fig. 3.20(c), the image charge produced by the conduction band effectively cancels out the dipole which then cannot interact with the incident electromagnetic field polarized in that direction.

The MSSR is obviously very useful for determining the symmetry of adsorbate complexes because for a given symmetry, only certain vibrational modes are expected to appear.

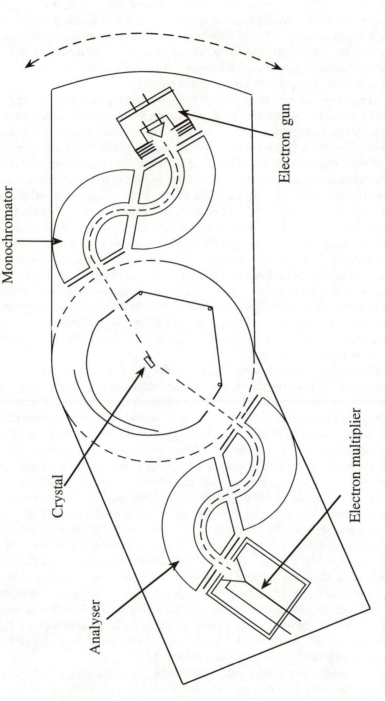

Fig. 3.18 The EELS analyser. Electrons are emitted from a hot filament in the electron gun and pass into a monochromator to select out electrons which have only certain energies within a small range. After leaving the monochromator, the electrons are incident on the sample and are scattered into an energy analyser. The energy analyser is swept across the entire energy range and peaks occur in the recorded spectrum at the energies at which they have lost or gained energy.

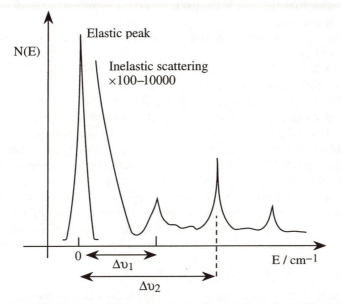

Fig. 3.19 A schematic EELS spectrum, plotted in the form $N(E)$ against E. The inelastically scattered electrons appear at specific energies that are equivalent to vibrational quanta of energy (e.g. Δv_1 and Δv_2 in the figure).

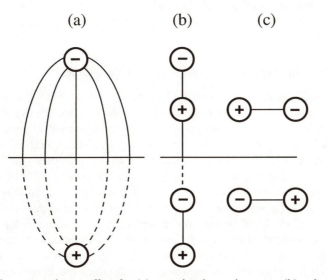

Fig. 3.20 The image charge effect for (a) a single, charged species, (b) a dipole which is perpendicular to the surface, causing an enhancement by a factor of 2, and (c) a dipole parallel to the surface, which causes the effective cancellation of the charge.

It is interesting to note that the electrons which are scattered by the dipole scattering mechanism do not leave the surface at quite the same angle to the surface normal as that for geometrical specular reflection. They are scattered at a slightly smaller angle to the surface normal ($\sim$1–2°) as shown in Fig. 3.21(a), in the so-called dipolar scattering lobe.

The reason for this is that momentum parallel to the surface must be conserved. The array of adsorbed molecules oscillate and thereby create an electric field, which interacts with the incoming electrons. When the adsorbates oscillate in phase, their electric field is at its strongest and so they couple strongly with the incoming electrons. This can be thought of as a surface wave with a long wavelength, and momentum parallel to the surface, $Q_\parallel$, which is small and finite, as shown in Fig. 3.21(b). If $Q_\parallel$ is zero then the beam is elastically scattered in the specular direction. However, in the case of dipole scattering, where the interaction is strong and the energy transfer $\Delta E = h\nu_{\text{vibrational}}$ is small but not negligible in comparison to the initial beam energy E_p, the effect is to produce a small decrease in scattering angle. Thus momentum is conserved, since the initial parallel momentum $k_{i\parallel}$ is equal to the dipole scattered parallel momentum, $k_{f\parallel}$, plus the parallel momentum of the surface wave, $Q_\parallel$, that is

$$k_{i\parallel} = k_{f\parallel} + Q_\parallel \tag{3.56}$$

There are other scattering mechanisms for VEELS. Impact scattering is a short-range interaction based on collisional interactions of the electrons with the surface species; it too has selection rules for on-specular reflection, which are different to those of dipole scattering. Additional peaks are found when the spectra are recorded off-specular and these measurements are also found to be very useful for determining adsorbate structures.

Another scattering mechanism is that of negative ion resonances, where the incident electrons can become trapped in the molecular orbitals. Once trapped, they take on the characteristics of the orbitals and are ejected from the orbitals with an angular distribution and selection rules which are characteristic of the orbital in which they were trapped.

3.5.1.2 *Reflection–absorption infrared spectroscopy (RAIRS)*

The use of infrared spectroscopy has certain advantages over VEELS but is technically more difficult because of the sensitivity problems described earlier. The IR spectrum of an adsorbate is recorded by taking a single reflection of an infrared beam at grazing angles of incidence and taking the ratio of the resulting spectrum to the spectrum obtained for reflection off the clean surface. The presence of the adsorbate causes small changes in reflectivity, ΔR, at the frequencies of the adsorbate vibrations, where the IR is absorbed. The spectrum is therefore produced in the form $\Delta R/R$ (where R is the total reflectivity of the surface) against frequency. Grazing angles are required because this is the condition under which the reflectivity of IR off a metal surface is at a minimum for p-polarized radiation and this leads to absorption by the modes with vibrational dipole changes perpendicular to the surface. RAIRS spectra are hence also subject to the metal surface selection rule. In this case the electromagnetic field with which the dipoles interact is that of the IR radiation (rather than that due to the incident electrons). RAIRS spectra are therefore directly comparable to those from the dipolar excitation VEELS.

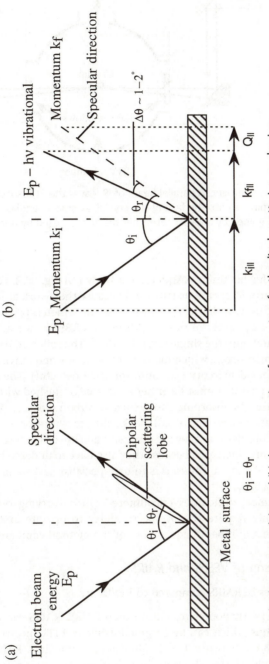

Fig. 3.21 (a) The dipolar scattering lobe and (b) the conservation of momentum for the dipole scattering mechanism.

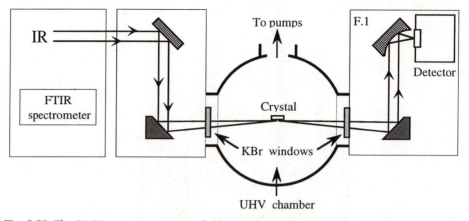

Fig. 3.22 The RAIRS experiment. A parallel beam of IR leaves the FTIR spectrometer and is focussed by a set of mirrors onto the face of the crystal, at grazing angles of incidence. The scattered radiation is guided by another set of mirrors and focussed by mirror F.1 onto the detector.

A typical system for obtaining RAIRS data is shown in Fig. 3.22. Fourier transform infrared (FTIR) spectrometers are usually used because they work at higher sensitivity than dispersive instruments. In the figure the IR beam (which is produced by a 'globar' source) is directed via a series of mirrors, through a KBr or an NaCl window into a vacuum chamber, and onto the single-crystal surface. The reflected beam passes out of the chamber and onto mirrors which focus it onto a photoconductive detector (usually a liquid nitrogen cooled mercury cadmium telluride detector). The optical pathway outside the vacuum chamber must be either evacuated or flushed with nitrogen or dry air containing no other IR absorbing gases, such as carbon dioxide, as the absorptions of these species can swamp the small peaks due to the adsorbate. The window materials can cause problems as they are quite soft and are difficult to seal to ultrahigh vacuum. The problem is increased because they also have to undergo bake-out at about 125°C for 48 hours. The usual solution is to use specialist gaskets made of viton (see Chapter 2, Section 2.5.1).

Despite the technical difficulties encountered when carrying out RAIRS experiments, which usually require the IR spectrometer to operate at the limits of its sensitivity, there are a number of advantages of the method compared with VEELS.

3.5.1.3 A comparison of VEELS and RAIRS

The main advantages of RAIRS compared to VEELS are:

(i) VEELS has inherently poorer resolution than RAIRS. RAIRS can be performed at all the resolutions which can be obtained using an FTIR spectrometer (typically the range 0.5–8 cm^{-1} is used). This often means that in a RAIR spectrum of a complex molecule, all the vibrations in a given spectral region can be resolved; in VEELS (where the 'usual' resolution is 20–80 cm^{-1} as we have seen, but with 8 cm^{-1} as an attainable future standard), they often all overlap into broad bands.

(ii) RAIRS can be used under much higher pressures than VEELS. VEELS is restricted to use in vacua of about 10^{-6} mbar, because a coherent electron beam cannot be maintained at higher pressures. RAIRS, on the other hand, can be operated even at pressures above atmospheric pressure, provided the windows are properly sealed on. This offers the possibility of using RAIRS data to correlate more directly with the results obtained for transmission IR studies of real catalyst materials.

(iii) High wavenumber absorptions are comparatively strong.

This does not mean that VEELS has been entirely replaced by the higher resolution RAIRS experiments, even though they have similar sensitivities. There is still much additional valuable information to be gained from VEELS.

The main advantages of VEELS compared to RAIRS are:

(i) Impact excited modes and negative ion resonances provide valuable additional information on non-completely-symmetrical vibrations.

(ii) The study of vibrational modes in the low-energy region of the IR is very important because it is in this region that the metal–adsorbate vibrations appear, and those originating from frustrated translations and rotations. The spectral range of RAIRS is limited in this region by window materials (such as NaCl which has a lower limit of 600 cm^{-1} and KBr which has a lower limit of 400 cm^{-1}), conventional globar sources which have low output below 600 cm^{-1}, and detectors which often exhibit inadequate sensitivity in the far-IR region. The spectral region from 100 to 400 cm^{-1} is therefore, as yet, difficult to investigate with laboratory-based RAIRS but can be studied using synchrotron radiation sources and helium-cooled bolometer detectors. With VEELS, on the other hand, studying this region is routine.

(iii) Low-wavenumber absorptions are comparatively strong.

VEELS and RAIRS can therefore be considered as complementary techniques. Their use for elucidating molecular adsorption phenomena will be demonstrated in the next sections.

3.5.2 CO adsorption on metal surfaces

The adsorption of CO on surfaces is very well studied for a number of reasons. First, it is a simple adsorbate which is relatively easy to model theoretically, in terms of modelling both spectra and the electronic structure. Depending on the nature of the substrate, CO can physisorb or it can chemisorb both non-dissociatively and dissociatively. It has several orbitals that can take part in bonding to the surface. In addition, it can occupy different adsorption sites on the surfaces of different metals and, from a more practical point of view, it is important in a large number of catalytic processes, such as the synthesis of methanol.

3.5.2.1 *The nature of the adsorption process: UPES and XPES measurements*

The first experimental evidence regarding the nature of the binding of CO to metal surfaces came from the use of photoelectron spectroscopy (PES) methods (especially

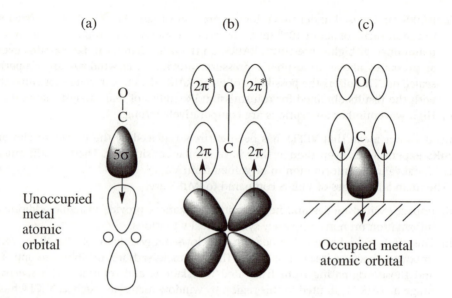

Fig. 3.23 A schematic diagram of synergic bonding of CO to a metal. The interaction of (a) the 5σ orbital and (b) the $2\pi^*$ orbital with a metal in a metal carbonyl compound. (c) The interaction of adsorbed CO with a surface.

UPES) to look at the interactions of the molecular orbitals with the metal orbitals of the substrate. In CO the molecular orbitals are

$$1\sigma^2 \ 2\sigma^2 \ 3\sigma^2 \ 4\sigma^2 \ 1\pi^4 \ 5\sigma^2 \ 2\pi^*$$

The 4σ orbital is localized on the oxygen atom while the 5σ orbital is localized on the carbon atom and both of these orbitals are non-bonding. The empty $2\pi^*$ antibonding orbital is also available to take part in the interaction with the surface. When CO chemisorbs molecularly it is found to bind via the carbon lone pair in the 5σ orbital, just as it does when CO as a ligand bonds to the metal in metal carbonyl compounds. The orbitals involved are shown in Fig. 3.23 for a metal carbonyl. This combination of σ and π orbitals of CO in the interaction with the surface is called synergic bonding. A covalent bond is formed by donation of electrons from the 5σ orbital to a vacant metal d orbital, see Fig. 3.23(a). At the same time, the full d orbitals are able to donate electron density into the vacant $2\pi^*$ orbitals, as illustrated in Fig. 3.23(b). On adsorption, the situation is analogous, as shown in Fig. 3.23(c). The effect on the energies of the molecular orbitals of CO adsorbing on a metal surface is apparent from the UPES spectra. Figure 3.24 shows schematically what is observed by UPES for gas-phase CO and for CO adsorbed on a surface. On adsorption, the 5σ orbital shifts down in energy, which confirms that the bonding is via the carbon atom to the surface; the 5σ orbital essentially becomes a bonding orbital.

Obviously, the strength of interaction with the surface and the interatomic CO bond distance of the adsorbed CO are highly dependent on the amount of electron transfer between the 5σ and $2\pi^*$ orbitals with the surface. A delicate balance of forces results which is highly dependent on the substrate itself and also affects the 4σ and 1π orbitals.

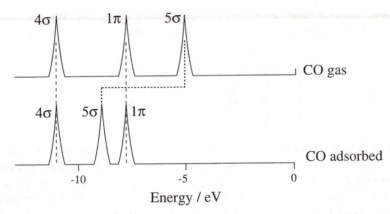

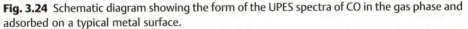

Fig. 3.24 Schematic diagram showing the form of the UPES spectra of CO in the gas phase and adsorbed on a typical metal surface.

For example, backdonation of electrons into the $2\pi^*$ orbital 'weakens' the CO bond and allows it to stretch. This increase in bond length raises the energy of both 4σ and 1π orbitals, with the latter being raised more than the former.

The strength of chemisorption is most reflected, however, in the relative positions of the 5σ and 1π orbitals which become closer in energy as the strength of bonding increases. For strong chemisorption they can overlap and the two UPES peaks can merge.

For physisorbed CO there is little difference in UPES spectra for the adsorbed species compared to the spectra in the gas phase and the interaction is a weak van der Waals one.

XPES is also of value for studying the CO adsorption process as it can be used to distinguish between dissociative and non-dissociative chemisorption. As we saw in Chapter 2, XPES is element specific because it mainly probes the core levels of atoms and there are relatively few overlaps between elements. It turns out that the ionization potential of an atomic core level has some dependence on the chemical environment in which it is sited and so the electrons which are ejected have an energy which is shifted from the gas phase value by this so-called 'chemical shift' (of up to about 10 eV). This chemical shift can be of considerable value in determining the nature of bonding between adsorbate and surface. For example, if one studies the peak resulting from the C(1s) level, it is found that its position for molecularly bound CO is very different from that of C(1s) in graphitic and carbidic forms of carbon; these forms can also be readily distinguished by XPES.

It is common, therefore, to study the adsorption of CO using a combination of UPES and XPES to determine the nature of the interaction with the surface. Both techniques are usually used to study the interaction starting at low temperature, through the temperature range until desorption or decomposition takes place.

In surface science, as noted in the introduction to this section (3.5), a combination of techniques is needed to gain a good understanding of the adsorption system under consideration. Correlation of the PES data with the heats of adsorption of CO is found to be very valuable. It is found that if $\Delta H_{ads} > 260$ kJ mol^{-1} dissociative adsorption tends to result rather than molecular chemisorption which occurs at lower values. The distinction between dissociative and non-dissociative chemisorption also correlates systematically with the position of the substrate in the periodic table as shown below.

Adsorption of CO

Dissociative			Non-dissociative		
Cr	Mn	Fe	Co	Ni	Cu
Mo	Tc	Ru	Rh	Pd	Ag
W	Re	Os	Ir	Pt	Au

Dissociative chemisorption tends to occur for adsorption on substrates on the left of the periodic table and non-dissociative chemisorption occurs on substrates on the right. For metals on the right of the periodic table the Fermi energy E_f is below the level of the highest energy $2\pi^*$ orbital of the CO, so there is little electron donation into it and the CO bond is strong. On the left-hand side, E_f is above the level of the $2\pi^*$ CO orbital and so substantial electron donation can occur, weakening the CO bond and stabilizing the M–C bond, which then enables the CO to dissociate.

3.5.2.2 *Adsorption site and geometry: vibrational spectroscopy*

In Section 3.5.1 the use of vibrational spectroscopy to study adsorption was described, and here it is applied to the adsorption of CO to gain information on adsorption site and geometry.

The strength of the interaction of CO with the surface is reflected in the vibrational frequency of the CO bond-stretching vibration. The stronger the CO bond, the higher the frequency and the weaker the adsorption on the surface.

VEELS was the first technique to be used to study the vibrational spectra of CO adsorbed on single-crystal surfaces and an example is shown in Fig. 3.25 for CO chemisorbed on Pt(111) at maximum (saturation) coverage. The spectrum was obtained with one of the new generation of high-resolution spectrometers and the resolution obtained was about 3.4 cm^{-1}. Recalling that for dipole scattering in VEELS, the MSSR states that only modes perpendicular to the surface are allowed, it is immediately apparent that the molecule does not lie parallel to the surface; in which case no CO stretching mode would be observed. There are five peaks in the spectrum, at 402, 468, 1768, 1846 and 2088 cm^{-1}. At lower coverage, only two modes at 468 and 2088 cm^{-1} are found, so this implies that there are two types of CO on the surface at high coverage. We know from UPES that the CO is bonded via the carbon atom to the surface and the VEELS results support this. The peaks in the VEELS spectra are assigned by analogy with the IR spectra of metal carbonyls using characteristic group frequencies.

The two peaks at 468 and 2088 cm^{-1} are therefore assigned to the Pt–C and C–O stretching modes (ν_{PtC} and ν_{CO}) of a linear species Pt–C≡O. On the surface, this is described as a terminally bound CO (or 'on-top'), where the CO is bound to only one Pt atom and is oriented perpendicular to the surface. The peaks at 402, and 1846 cm^{-1} are assigned to the Pt–C and C–O stretching modes of CO bridging two platinum atoms (a μ_2-bridging species); again the CO direction is perpendicular to the surface. The 1768 cm^{-1} peak which appears as a 'shoulder' on the side of the 1846 cm^{-1} peak is due to the CO stretching mode which results when some of the CO molecules bridge across three

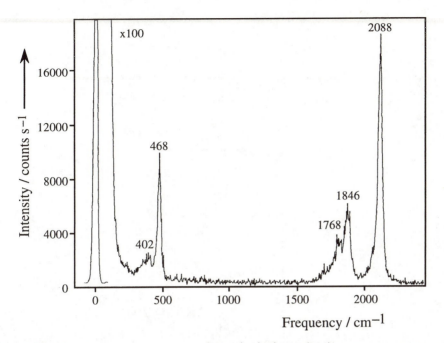

Fig. 3.25 EELS of a saturation coverage of CO adsorbed on Pt(111) at room temperature. The electron beam energy was 3 eV. (Spectrum recorded by Dr E. Hargreaves and Prof. M.A. Chesters of the Department of Chemistry, University of Nottingham.)

metal atoms in a μ_3-bridging site. The CO stretching frequency turns out to be more useful for determining the adsorption site than does the MC stretching frequency, as there is greater overlap of the frequency ranges for the MC stretches of terminal μ_2- or μ_3-bridging CO species on different metal substrates. The MC stretching modes for μ_2- and μ_3-bridging species are thought be very close in frequency in the Pt(111)/CO case and so cannot be separated. Figure 3.26 shows the bonding sites and expected frequency ranges in wavenumber units for the ν_{CO} stretching modes.

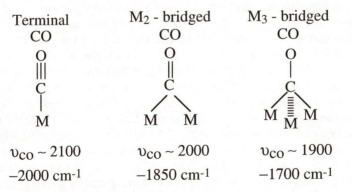

Fig. 3.26 Typical frequency ranges for the CO stretching mode for carbon monoxide absorbed in terminal, μ_2- and μ_3-sites.

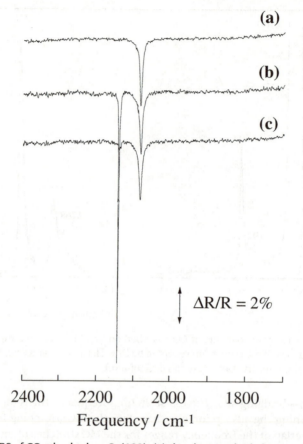

Fig. 3.27 RAIRS of CO adsorbed on Cu(100). (a) the chemisorbed phase at 23 K, (b) further CO adsorption to give a chemisorbed layer plus physisorbed multilayers and (c) the surface in (b) following heating to 26 K. (From Camplin, Cook and McCash 1995.)

CO readily chemisorbs on many metal surfaces in the 'usual' temperature range of surface science experiments (this usual range is generally from about 77 K to the anneal temperature, see Section 2.5.2). By using a liquid helium cooled manipulator it is possible to observe physisorbed CO. An example of an RAIR spectrum of CO on Cu(100) at 23 K is shown in Fig. 3.27. In Fig. 3.27(a) the CO chemisorbs in the first layer with a characteristic CO stretching frequency, v_{CO} of 2086 cm^{-1} indicating that the CO is terminally bound. In Fig. 3.27(b) physisorbed layers are seen to grow in above the initial chemisorbed ones with a characteristic v_{CO} of 2143 cm^{-1}. This is close to the value for gas-phase CO and indicates the weak interaction of the CO with the surface. If the surface is heated to 26 K as shown in Fig. 3.27(c), the multilayers of physisorbed CO desorb and leave only a small peak at 2143 cm^{-1}. This is due to a single physisorbed layer, which is desorbed above 35 K. This difference in desorption temperature between the first and subsequent physisorbed layers indicates that the first layer is slightly more strongly bound to the surface than are the other layers, for reasons noted in Section 3.3.2.

3.5.2.3 *The work function change on adsorption of CO*

The work function change observed on adsorption of CO is dependent on the precise interaction with the specific substrate. For example, on Rh(111), CO occupies bridging sites and behaves as expected for an electronegative adsorbate so the work function increases with coverage. On the other hand, on Pt(111) the work function is initially found to decrease to -0.5 eV at a coverage of $\theta' = 1/3$, and then to increase to zero again at the maximum coverage of $\theta' = 1/2$. These Pt results are explained by considering the transfer of charge between adsorbate and surface. In the initial adsorption phase, when the CO is terminally bound, it behaves as an electropositive adsorbate, donating electron density to the surface from the 5σ orbitals. At higher coverages, CO–CO repulsion prevents every terminal site from becoming occupied and so the less favourable bridged sites become populated. Charge transfer for the bridged CO is dominated by backbonding to the $2\pi^*$ orbitals and so the bridged CO effectively acts as an electronegative adsorbate. At maximum coverage the electropositive contributions from the terminally bound CO and the electronegative contributions from the bridged CO roughly balance and so the overall work-function change is zero.

3.5.2.4 *Structural changes: LEED*

In Section 2.6.2.2 the use of diffraction methods to study surface structure was described. LEED is of particular value for the study of adsorption phenomena. Following adsorption, the diffraction pattern obtained from a surface is often radically altered as new, additional spots appear. These are due usually to the overlayer of the adsorbate on the surface. LEED cannot be readily used to determine the adsorption sites occupied, with the exception of cases when a complete dynamical calculation can be carried out. However, it can be used to determine the structure and periodicity of the overlayer which forms on the surface.

In order to make use of LEED, we must consider the manner in which the overlayer patterns can be described and this is generally done using either matrices or Wood's notation. Because of its common usage, we will use Wood's notation. The unit vectors of the overlayer $\underline{a}'$ and $\underline{b}'$ are described by vectors parallel to the vectors of the substrate spots $\underline{a}$ and $\underline{b}$, that is

$$\underline{a}' = n\underline{a} \quad \text{and} \quad \underline{b}' = m\underline{b} \tag{3.57}$$

n and m are integers which describe the primitive unit cell $p\,(n \times m)$. The index for the overlayer is then written

$$M\,(h\,k\,1)\,(n \times m) \tag{3.58}$$

where M is the metal substrate and $(h\,k\,l)$ is the face of the crystal being studied. In a complete description the adsorbate and coverage are sometimes quoted at the end of the index—but since it is often difficult to measure the exact coverage of the overlayer this information may be omitted.

It is often found that the vectors of the unit mesh for the overlayer do not follow along the same directions as the vectors for the substrate. In this case the angle by which they are rotated ($R\theta°$) is also given:

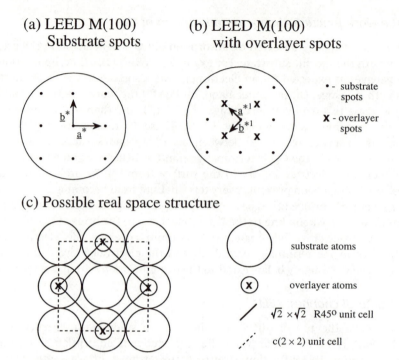

(a) LEED M(100)
Substrate spots

(b) LEED M(100)
with overlayer spots

• - substrate spots

X - overlayer spots

(c) Possible real space structure

substrate atoms

Ⓧ overlayer atoms

$\sqrt{2} \times \sqrt{2}$ R45° unit cell

c(2 × 2) unit cell

Fig. 3.28 (a) The LEED pattern for a clean M(100) face of an fcc metal, indicating the two unit vectors $\underline{a}^*$ and $\underline{b}^*$. (b) The LEED pattern for the same (100) face with overlayer spots at the half-order positions, with unit vectors $\underline{a}^{*1}$ and $\underline{b}^{*1}$ and (c) A possible real space mesh which could explain the pattern in (b). This is the $(\sqrt{2} \times \sqrt{2})$ R45° unit cell, which is the same mesh as that which is sometimes called the c(2 × 2). The unit cells for both of these are drawn on the figure.

$$M(h\,k\,1)\,(n \times m)\,R\,\theta^\circ \tag{3.59}$$

Rotation of 45° is quite common. In these cases it is possible that the literature will describe the overlayer as being centred. Figure 3.28(a) shows the LEED pattern for a clean M(100) surface. The LEED pattern in Fig. 3.28(b) is at the same beam energy so that all the substrate spots are in the same positions. The overlayer which has been adsorbed on the surface gives rise to the overlayer spots at the half-order positions as indicated. In Fig. 3.28(c), a suggested surface structure is shown which could reproduce the LEED pattern observed in (b). This can be described either as a $(\sqrt{2} \times \sqrt{2})$ R45° unit mesh (the primitive unit mesh, shown as a solid line) or as its equivalent surface structure, the centred c(2 × 2), which is a non-primitive unit mesh (shown as a dashed line).

LEED data can be used in conjunction with the other techniques described above to come up with a likely arrangement of the molecules on the surface. Returning to the CO on Cu(100) system which we looked at in terms of RAIRS in Section 3.5.2.2, we can use LEED and RAIRS together to predict the adsorbate positions on the surface. Two LEED patterns are observed for the Cu(100)/CO system. At half monolayer coverage

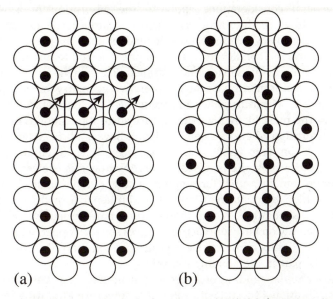

Fig. 3.29 The real space overlayer structures for CO on Cu(100). (a) the open $(\sqrt{2} \times \sqrt{2})$ R45° which is found at half monolayer coverage and (b) the compressed $(7\sqrt{2} \times 7\sqrt{2})$ R45° structure at higher coverage. The solid boxes indicate the unit cells and the arrows in (a) indicate the shift of the overlayer molecules to form the overlayer structure in (b).

$(\theta' = 0.5)$, a $(\sqrt{2} \times \sqrt{2})$ R45° overlayer is formed and when further CO is adsorbed, a $(7\sqrt{2} \times \sqrt{2})$ R45° structure is observed at the highest coverage obtained which is at $\theta' = 0.57$. Bearing in mind the RAIRS data which show that CO adopts only terminal sites on the surface, the predicted arrangement of CO molecules in the $(\sqrt{2} \times \sqrt{2})$ R45° structure is shown in Fig. 3.29(a).

At high coverages there is insufficient room on the surface to accommodate further CO atoms perpendicular to the surface and so the layer has to 'compress' to form the $(7\sqrt{2} \times \sqrt{2})$ R45° structure. As there is no evidence from RAIRS of CO adopting bridged sites, this compression must involve some tilting of some of the COs and the predicted arrangement is shown in Fig. 3.29(b).

3.5.3 Adsorption sites and bonding to metal surfaces by hydrocarbon molecules

The adsorption of hydrocarbon molecules is of intense interest to surface scientists because of their importance in industrial catalytic processes. There are many processes involving hydrocarbons, such as hydrogenation/dehydrogenation, hydrogenolysis, isomerization, increasing and decreasing the carbon chain length, and so on. An understanding of the manner in which they bind to the surface is important, because it provides important clues as to the mechanisms by which these processes proceed. In this section we look at some examples of hydrocarbon adsorption, largely using vibrational spectroscopy to investigate the adsorption process.

A few general considerations apply to the adsorption of a wide range of hydrocarbons. As for the interactions of CO with a surface, the interactions of a given hydrocarbon molecule with a surface are dependent upon the natures of the adsorbate and substrate, and the conditions (mainly of pressure and temperature) under which the adsorption process takes place (and is studied). The specific face of the substrate can also be important as it might not offer the preferred geometry for adsorption of a given surface complex. Physisorption can occur at low temperatures. Above these temperatures the most likely low-temperature adsorption state to be encountered is that of chemisorption where C–C and C–H bonds have not broken (i.e. non-dissociative chemisorption). On reasonably reactive surfaces, it is often found that as the temperature of the substrate is raised further, hydrogen is evolved from the surface and the carbon remains, often forming ultimately a graphitic or carbidic layer. This implies that at high temperatures CH bonds are broken to form dissociatively adsorbed surface species. Generally, at room temperature, non-dissociative molecular adsorption is observed only on fairly unreactive surfaces such as copper, while fragmentation of the molecules occurs on more reactive surfaces with partially filled d-bands which are able to take part in the bonding process.

Much of our understanding of hydrocarbon interactions with surfaces has been gained from vibrational spectra with additional 'fingerprint' information derived from the patterns observed in electronic spectra such as UPES and Auger electron spectroscopy (see Section 3.6.1).

Group frequency information is vital in the interpretation of the hydrocarbon spectra. A useful guide to positions of the expected modes comes from the frequencies expected for the C–C and the completely symmetrical C–H vibrations of the molecular species, ethane, ethene and ethyne.

Molecule	ν_{C-C}/cm^{-1}	ν_{C-H}/cm^{-1}
Ethane	990	2870
Ethene	1620	3020
Ethyne	1974	3374

3.5.3.1 *Saturated hydrocarbons*

The interactions of these highly energetically stable compounds with surfaces tend to be weak; physisorbed layers are usually found and substantial coverages are only produced at low temperatures. Figure 3.30 shows the RAIR spectrum of ethane adsorbed on Cu(111) at 91 K. The figure also illustrates the vibrational modes which are observed. The spectrum is dominated by the asymmetric CH_3 stretching mode at 2961 cm^{-1} and a moderately intense broad band occurs at 1458 cm^{-1}, which is due to the asymmetric deformation. All of the other features of the spectrum are very weak. The modes are all about 30 cm^{-1} lower in frequency than for the gas-phase molecule, which shows that the molecules in the first layer are interacting with the surface, but that this interaction is fairly weak. This confirms that they are, indeed, physisorbed. Application of the MSSR to the observed spectrum indicates that the intense CH_3 asymmetric stretching and asymmetric deformation modes, and the low intensity of the other modes, are due to a bonding geometry where the molecules lie with the C–C axis parallel to the surface. If the molecules were perpendicular to the surface, a symmetric

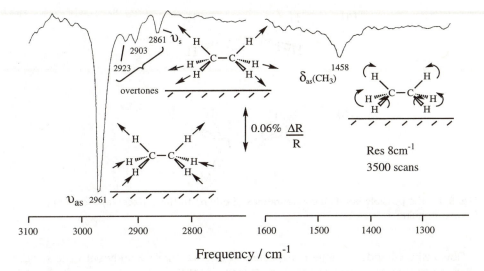

Fig. 3.30 The RAIRS spectrum of ethane adsorbed on Cu(111) at 91 K. The eclipsed bonding configuration and its observed vibrational modes are also shown. (This spectrum was recorded by the author).

methyl stretching mode would dominate the spectrum. The binding configuration with the CH_3 groups in either staggered or eclipsed positions is possible. The modes are illustrated in Fig. 3.30 for the eclipsed conformation.

Longer chain and cyclic alkanes are also found to physisorb, with the carbon skeleton parallel, or as parallel as possible, to the surface.

3.5.3.2 *The alkenes*

The molecular non-dissociative adsorption of ethene bears some similarity to that of ethane, in that ethene also tends to adsorb with the CC bond parallel to the surface. However, the surface–adsorbate interaction is usually stronger for ethene compared to ethane, and the bonding is very different. In terms of bonding, the adsorption is achieved in one of two ways, either by π-bonding between the C=C double bond and the surface, as shown in Fig. 3.31(a), or by di-σ-bonding from the two carbon atoms to different metal atoms in the surface, as shown in Fig. 3.31(b).

The RAIR spectrum for ethene π-bonded to Cu(111) is shown in Fig. 3.32, with the observed vibrational modes indicated. The frequencies of the peaks are close to those of the free ethene molecule. The only intense mode in the spectrum is at 990 cm^{-1} which is due to the symmetric out-of-plane CH_2 deformation. The other modes are very weak because the molecule is nearly planar and parallel to the surface. The symmetric CH_2 stretch and the symmetric CH_2 deformation are weakly observed because the CHs are tilted slightly away from the surface. The C=C stretching mode is also weakly observed because a small dipole moment perpendicular to the surface is produced during the vibration. (Note that CC vibrations for species where the CC bond lies parallel to the surface are quite commonly observed. They are far more obvious for some ethyne interactions with surfaces and will be described in more detail in the next section.)

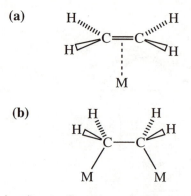

Fig. 3.31 The possible bonding configurations of ethene to the surface: (a) π-bonding and (b) di-σ-bonding.

Obviously, π-bonded ethene is expected to interact more weakly with the surface than the di-σ species, such as occurs on Pt(111) at 77 K. It is relatively easy to tell the difference between the two species vibrationally because in the di-σ species, the CH bonds are tilted well away from parallel to the surface. In fact, the carbon atoms can be considered to be close to an sp^3 hybridized form. This means that not only do modes such as the symmetric CH_2 stretch and symmetric CH_2 deformation modes become more intense, but also that the frequencies are shifted from the values of the parent ethene molecule because of the stronger interaction with and different binding to the surface. In this way, vibrational spectroscopy not only helps us to determine the orientation of adsorbates but also gives us insights into the nature of the bonding to the surface.

At higher temperatures the di-σ species is often found to fragment and rearrange on the surface to form the ethylidyne CCH_3 species, which has the C–C axis perpendicular to the surface. A suggested scheme for how this conversion takes place is shown in Fig. 3.33, together with further dissociation steps where more H is lost to form

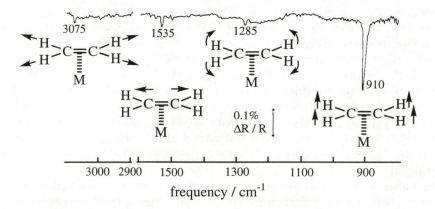

Fig. 3.32 RAIRS of the adsorption of ethene on Cu(111) with the associated vibrational modes. (From McCash 1990.)

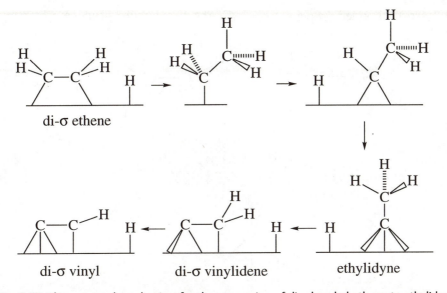

Fig. 3.33 The proposed mechanism for the conversion of di-σ-bonded ethene to ethylidyne, di-σ vinylidene and di-σ-bonded vinyl species on the Pt(111) surface.

vinylidyne (CCH_2) and sp^2 and sp^3 hybridized C_2H. Each of these species has been identified using vibrational spectroscopy.

As an example of how effective vibrational spectroscopy is for identifying these fragments, the RAIR spectrum of ethene adsorbed on Pt(111) at 360 K is shown in Fig. 3.34. The spectrum is radically different from that of the adsorbed molecule in its

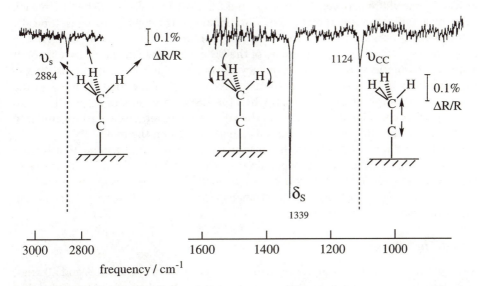

Fig. 3.34 RAIRS of ethylidyne adsorbed on Pt(111) at 360 K, with the observed vibrations. (From Chesters *et al.* 1990.)

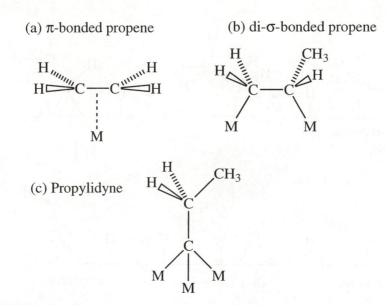

Fig. 3.35 The bonding of propene to surfaces, (a) π-bonded, (b) di-σ-bonded and (c) propylidyne.

di-σ form at 77 K. The only mode in the CH stretching region occurs at 2881 cm^{-1} and this is typical of a symmetric methyl stretching mode, indicating that the molecular fragment is bound with a methyl group perpendicular to the surface. This is confirmed by the observation of an intense symmetric methyl deformation at 1342 cm^{-1}. The CC stretch is also observed at 1124 cm^{-1} and from its position, shows that it has a bond order of somewhere between 1 and 2. CH and CC modes are distinguished from each other by adopting the usual method of vibrational spectroscopy, of using isotopically substituted adsorbates. In this case, the adsorption of C_2D_4 was used to confirm the assignments by shifting the frequencies of the CH modes down in energy by a factor of about $\sqrt{2}$, while leaving the CC frequency in roughly the same position as for the hydrogen-substituted analogue. The spectrum is closely similar to that obtained for the coordination complex $(CH_3C)Co_3(CO)_9$ when the MSSR-active modes perpendicular to the surface are considered, showing that the ethene has undergone a rearrangement to ethylidyne CCH_3. The observed vibrations are illustrated on the figure.

For heavier alkenes, bonding to the surface is similar to that of ethene, that is via π or di-σ bonding to the surface depending on the metal, as shown for propene in Fig. 3.35(a) and (b), Fragmentation occurs to give alkylidyne species as shown for propylidyne in Fig. 3.35(c). Again, vibrational spectra provide the key to determining their orientations on the surface.

3.5.3.3 *The alkynes*

As might be expected by comparison with the adsorption of the alkenes and alkanes, the interaction between the alkynes and the surface is usually via the CC bond and/or the carbons on either end of it. For ethyne, π-bonded species are found on some

(a) π-bonded

H—C≡C—H

M

(b) di-σ- + π- bonded (μ₃)

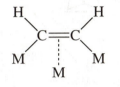

(c) di-σ-+di-π- bonded (μ₄)

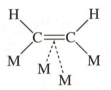

Fig. 3.36 Possible bonding configurations of ethyne on surfaces (a) π-bonded, (b) di-σ- + π-bonded, which is usually called μ_3-bound or Type B and (c) di-σ- + di-π-bonded, which is usually called μ_4-bound or Type A.

surfaces, such as Ag(111) as shown in Fig. 3.36(a), but it is more common to find much stronger bonding even if the surface is relatively unreactive, e.g. copper. The most commonly observed configurations are those of a μ_3-bonded molecule, where there is a di-σ interaction with two surface atoms and π-bonding to a third (so-called Type B) with the C_2H_2 plane tilted from parallel to the surface, and a μ_4-bonded molecule where there is a di-σ interaction with two surface atoms and di-π bonding to two others (so-called Type A) and the C_2H_2 plane is perpendicular to the surface. These bonding configurations are shown in Fig. 3.36: (b) illustrates Type B bonding and (c) Type A.

The 'on-specular' VEELS spectra of ethyne bound to Pd(110) at 100 K and to Cu(111) at 110 K are shown in Fig. 3.37 together with the vibrational modes to which the peaks are assigned. Ethyne binds as a Type B species on Pd(110) and as a Type A species on Cu(111).

For both Type A and Type B species, the interaction of the original C≡C bond with the metal atoms is sufficiently strong that the CH bonds are forced to tilt well away from the surface and the carbon orbitals are hybridized to somewhere between sp^2 and sp^3 in character. This is why both the CH stretch (at 2984 cm^{-1} on Pd and 2920 cm^{-1} on Cu) and CH in-plane deformation modes (at 913 cm^{-1} on Pd and 920 cm^{-1} on Cu) are observed, as they have dipole moments with components perpendicular to the surface.

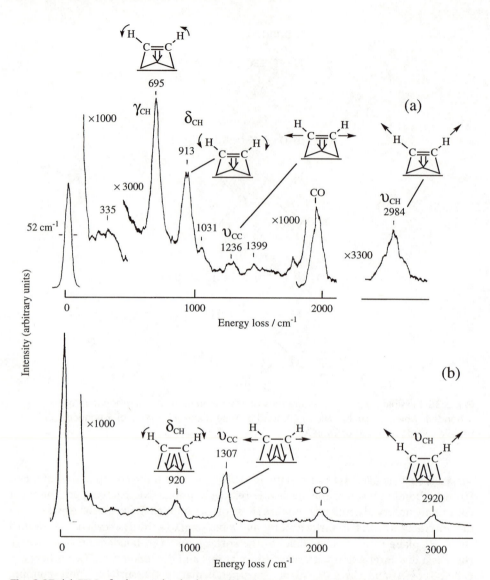

Fig. 3.37 (a) EELS of ethyne adsorbed on Pd(110) (Type B) at 100 K, with the observed vibrations. (Spectrum provided by Dr G.S. McDougall of the Department of Chemistry, University of Edinburgh, and Professor M.A. Chesters, Department of Chemistry, University of Nottingham.) (b) EELS of ethyne adsorbed on Cu(111) (Type A) at 110 K, with the observed vibrations. (From Bandy *et al.* 1984.)

They are also shifted considerably from the frequencies expected from the free molecule and are, vibrationally, much closer to those for ethane.

The most serious difficulty with the interpretation the spectra for ethyne molecules, which lie with the CC axis parallel to the surface, was that for the CC stretching mode to be allowed under the MSSR, it must have a dynamic dipole perpendicular to the

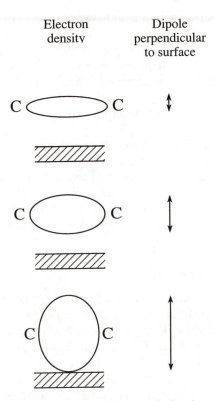

Electron density   Dipole perpendicular to surface

Fig. 3.38 Illustration of how the C≡C stretching mode for the ethyne species, where the CC axis lies parallel to the surface, produces a dipole perpendicular to the surface.

surface. At first sight it appeared that this was not the case. However, although the CC bond stretches parallel to the surface, it *does* produce an oscillating dipole perpendicular to the surface. This is because the degree of electron transfer between the CC bond and the surface varies substantially as a function of the fluctuating length of the vibrating bond as illustrated in Fig. 3.38.

The Type A μ_4 species which is formed on Cu(111) has a very strong CC stretching mode at 1307 cm^{-1}; the mode is also observed for the Type B μ_3 species on Pd(110), at 1236 cm^{-1}, but it is rather weak in comparison as Fig. 3.37 shows. The intensity of this mode illustrates the most striking difference between the Type A and B species as it illustrates dramatically the different strengths of interaction with the surface. The other main feature of the ethyne/Pd(110) spectrum is the appearance of the out-of-plane deformation at 695 cm^{-1} which, because of the operation of the MSSR, is not observed at all for the ethyne/Cu(111) system.

As in the alkene case, therefore, the vibrational spectra of adsorbed ethyne can be used to determine the nature of binding to the surface because the observed modes are characteristic of the type of interaction involved; each has a different 'signature' in terms of frequency and intensity in the vibrational spectra. In the case of Type A and Type B adsorbed ethynes, similar frequencies are expected but their relative intensities are very different.

On reactive surfaces, rearrangements and fragmentation can occur to produce similar adsorbates to those formed from the alkenes, especially if the surface is warmed in the presence of hydrogen. Disproportionation of adsorbed ethyne can occur to give $C_2H + C_2H_3$ and at high temperatures, sp^2 and sp^3 hybridized CCH species have been observed like those formed by the alkenes. Indeed, fragmentation is observed for ethyne adsorbed on the Pd(110) surface on heating to room temperature, but the effect of heating the ethyne-covered Cu(111) surface results in molecular desorption.

For molecular adsorption of the longer chain alkynes, similar interactions are observed to those for ethyne itself, with the CC bond parallel to the surface and the hydrogens and methyl groups tilted away from the surface. Similar fragmentation patterns are also seen. Once again, the vibrational spectroscopies, particularly RAIRS, which has sufficient resolution to resolve all the bands in the spectrum, are particularly valuable for determining the nature of the adsorbed state.

3.5.3.4 Aromatic hydrocarbons

Under normal ultrahigh vacuum conditions, when benzene adsorbs on a surface it is generally found that the molecule orients with the carbon ring parallel to the surface. The interaction is of π-bonding nature and an intense vibrational peak is observed due to the symmetrical out-of-plane CH deformation mode. The symmetric CH stretching modes, which are formally allowed by the MSSR even when the molecule is parallel to the surface, tend to be so weak that they cannot be observed in the spectrum. Any reorientation can thus be detected using vibrational spectroscopy as these modes increase in intensity as soon as the molecule is tilted away from the parallel direction.

There has been very little work on substituted benzenes except for methyl benzene (toluene) which also appears to adsorb non-dissociatively, parallel to the surface, at low temperatures. At higher temperatures it is found to dissociate to $-CH_2$ and C_6H_5-.

3.5.4 Adsorbate-induced reconstruction

At this point it is useful to note that adsorbates can often have a considerable effect on the structure of the surface. In Section 2.6.1 we saw that reconstruction of clean surfaces is undergone to minimize surface energies. Following adsorption, especially where there is a particularly strong adsorbate–substrate interaction, further reconstructions can occur, again driven by the need to minimize the surface energy. In the case of the Si(111) 7 × 7, the adsorption of hydrogen causes the surface to 'unreconstruct' and form a 1 × 1 structure again, as each Si bonds to an H atom and the unsatisfied valency of each of the dangling bonds is satisfied.

Figure 3.39 shows the effect of adsorbing oxygen on the Cu(100) surface, causing a reconstruction to a (2 × 1) structure. The effect of the oxygen is apparently to 'remove' a row of copper atoms, leaving a 'missing row' structure in order to attain the lowest possible surface energy. The unit cell is shown on the figure.

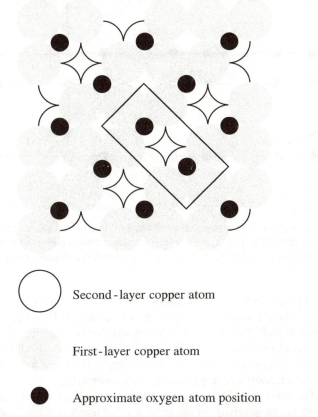

Second-layer copper atom

First-layer copper atom

Approximate oxygen atom position

Fig. 3.39 The Cu(100)/O reconstruction. The solid box indicates the (2 × 1) unit cell.

3.6 Surface chemical composition

Diffraction techniques and scanning tunnelling microscopy give us information on surface structure regularities and vibrational spectra can indicate the adsorption geometry and the type of bonding of the adsorbates on the surface, but none of these methods can provide definitive information on the chemical composition of surfaces. This is particularly important in surface chemistry and several techniques can be used. Two of the most common, Auger electron spectroscopy (AES) and secondary ion mass spectrometry (SIMS), are described below. Another important method which can be used to obtain this type of information is TPD (described in Section 3.4.2). In addition to the acquisition of kinetic data for desorbing molecules, TPD can also be used to observe molecular fragments as they desorb from a surface.

3.6.1 Auger electron spectroscopy (AES)

Auger electron spectroscopy (AES) is a process entirely complementary to that of X-ray fluorescence (XRF) which was described in Section 2.2.4.2. Figure 3.40 illustrates the

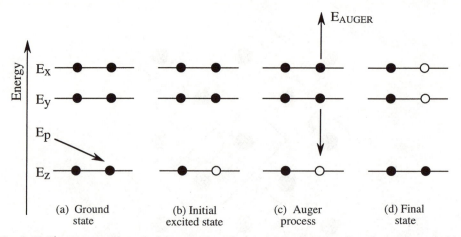

Fig. 3.40 The Auger process. In (a) excitation of a low-lying electron takes place. The unstable species formed (b) is stabilised by (c) an electron dropping from a higher level to fill the 'hole'. Another high-lying electron is emitted from the atom, leaving a doubly charged ion as in (d).

Auger process. In Fig. 3.40(a) an incident electron with high energy, E_p, displaces an electron which is in a low-lying energy level in the atom, E_x, in the same way as excitation occurs in XRF. This leaves a very unstable ionized species in Fig. 3.40(b). In order to 'relax', an electron from a higher level, E_y, drops down to fill the hole created in the orbital, releasing energy. The released energy is taken up by another electron, which is ejected from another energy level, E_z, as shown in Fig. 3.40(c), leaving a doubly charged final state shown in (d). E_{AUGER}, the energy with which the electron is released from the E_z level, is characteristic of the three energy levels involved and hence of the element from which it is produced. Equations (3.60) and (3.61) apply.

$$E_{AUGER} = E_x - E_y - E_z \qquad (3.60)$$

$$E_{AUGER} = E_{\text{Initial state}}^{(1+)} + E_{\text{Final State}}^{(2+)} \qquad (3.61)$$

From this we can see that the Auger energy is independent of the energy used for excitation. Auger and XRF are the two complementary pathways by which the initial excited state can relax. For processes where the initial electron is ejected from a very low-lying core level, XRF dominates the decay process. For initial low-energy excitation of the order of a few thousands of eV, no XRF is observed and all the decay is via the Auger process.

3.6.1.1 *Qualitative analysis*

Experimentally, the energy analysis is carried out in the same way as that for PES. In addition, the LEED system described in Section 2.6.2.2 can be used as a retarding field analyser (RFA) for the detection of the Auger electrons. A typical electron source consists of a hot tungsten filament with a repeller behind it set at -3 to -5 kV, which accelerates the electrons towards the sample.

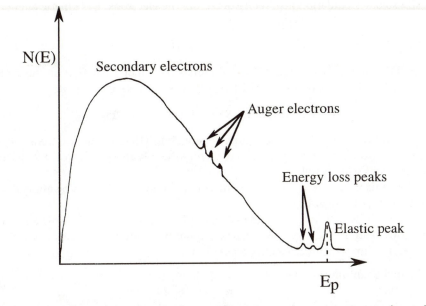

Fig. 3.41 Plot of *N(E)* against Energy, showing that Auger peaks are superimposed on a sloping secondary electron background. Energy E_p indicates the elastically scattered electrons.

The Auger peaks are sharp but are detected on a sloping background caused by the secondary electrons as described in Section 2.2.4.1. A typical spectrum is shown here in Fig. 3.41, with the AES peaks superimposed. In order to minimize the effect of the sloping background, the signal is differentiated, giving derivative peaks as shown in Fig. 3.42.

This makes it much easier to pick out the peak position, A. This is shown schematically in Fig. 3.42: (a) is the plot of *N(E)* against *E* and in (b) d*N(E)*/d*E* is plotted against *E*. The signal modulation is usually carried out electronically by applying a

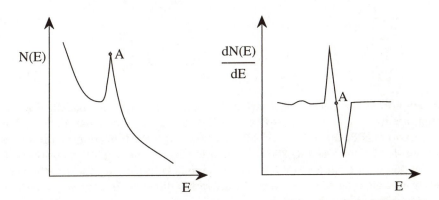

Fig. 3.42 (a) A plot of *N(E)* against *E* gives a peak on a highly sloping background as shown in figure 3.41. (b) after differentiation, a plot of d*N(E)*/d*E* against *E* gives a derivative peak on a flat background.

small sinusoidal modulation just before detecting the electrons. The kinetic energy of the collected electrons then takes on sinusoidal form

$$E + \Delta E \sin \omega t \qquad (3.62)$$

where E is the Auger energy and $\Delta E \sin \omega t$ defines the small oscillation on the signal. The number of electrons detected, $N(E + \Delta E)$, can be expanded in a Taylor series:

$$N(E + \Delta E) = N(E) + N'(E)\,\Delta E + N''(E)\,\Delta E^2/2! + \dots. \qquad (3.63)$$

The second term, $N''(E)\Delta E$, is an AC signal, dependent on the modulation frequency ω, and so if detection is carried out using a detector which is also tuned to ω, this part of the Taylor series is detected.

Note that this is a general type of modulation technique which can be applied to a wide range of spectroscopies.

The notation which can be adopted for AES spectra is that used for X-ray spectroscopy. Letters indicate the principal quantum number associated with the energy levels from which the electrons are ejected.

For principal quantum number, $n =$	1	2	3	4
X-ray notation	K	L	M	N

A subscript is used to denote the sublevel from which the electron is ejected. The lowest energy level in a given shell is denoted by a subscript 1, the next lowest by subscript 2, and so on. The energies of these levels are found from the atomic terms (using the l and j quantum numbers), and are related to the X-ray notation in the following way, starting with the lowest energy states and ending with the highest.

Quantum numbers			Atomic	X-ray
n	l	j	term	notation
1	0	1/2	1s	K
2	0	1/2	2s	L_1
2	1	1/2	$2p_{1/2}$	$\left.\right\}L_{2,3}$
2	1	3/2	$2p_{3/2}$	
3	0	1/2	3s	M_1
3	1	1/2	$3p_{1/2}$	$\left.\right\}M_{2,3}$
3	1	3/2	$3p_{3/2}$	
3	1	3/2	$3d_{3/2}$	$\left.\right\}M_{4,5}$
3	1	5/2	$3d_{5/2}$	

Two typical Auger spectra are shown in Fig. 3.43 for (a) Cr, which has the electronic configuration $1s^2\ 2s^2\ 2p^6\ 3s^2\ 3p^6\ 4s^2\ 3d^4$, and (b) Fe, which has the electronic configuration $1s^2\ 2s^2\ 2p^6\ 3s^2\ 3p^6\ 4s^2\ 3d^6$. In the example the transitions in the 400–600 eV region for Cr and the 550–750 eV region for Fe are due to LMM transitions. The peaks at 36 eV for Cr and 47 eV for Fe are due to LLM transitions. Clearly, the energy levels from which the AES electrons arise are rather different for the two, even though they only differ in atomic number by 2.

As the energies of the AES electrons evolved depend on three energy levels within the element, it can be seen that they are very specific to a given element. Different

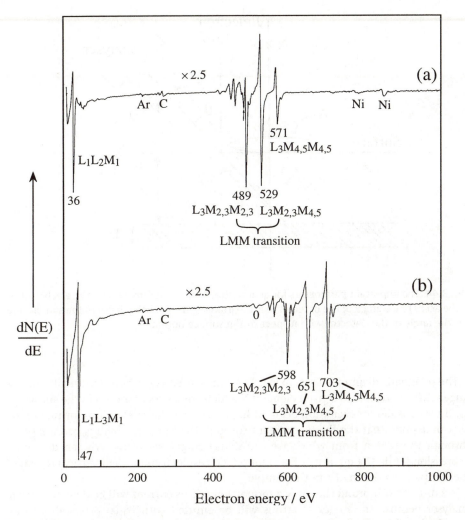

Fig. 3.43 AES spectra of (a) Cr and (b) Fe with the transitions involved indicated on the figure. (From Palmberg *et al.* 1972.)

elements in the same sample can therefore be easily distinguished, as they rarely have the same energy levels, and although some AES peaks from different elements may lie in overlapping energy regions, their positions are characteristic and independent of the excitation energy.

3.6.1.2 Quantitative analysis

AES spectra can be used to obtain quantitative information about the surface and near-surface regions. It should be noted that PES can be used as a quantitative tool in a similar way.

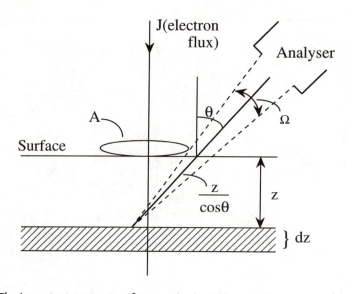

Fig. 3.44 The important parameters for quantitative AES measurements. J is the electron flux; Ω the acceptance angle of the detector; A is the irradiation area; z the depth of the sample and θ is the angle of the detector with respect to the surface normal.

The schematic diagram shown in Fig. 3.44 can be used as the basis for calculating the Auger current, i_A, due to a given element X which has a concentration in the surface of C_X. If we consider an excitation beam of high-energy incident electrons (or radiation), we can assume that the flux is constant through the sample, which is certainly true through the region from which we can detect Auger electrons. We can derive an expression for di, the quantity of current which results from a small vertical region of the sample, at a depth of z in the sample.

We must bear in mind that only a proportion of this current will be detected by the analyser because the Auger electrons will be emitted with equal probability in all directions (with a total solid angle of 4π rad) and the acceptance angle of the detector, Ω, is limited; thus the probability of the Auger electrons entering the detector is $4\pi/\Omega$. In addition, the spectrometer will not be 100 per cent efficient and so its transmission factor $G_{(E)}$ must also be taken into account. The signal is also attenuated by the effect of losses which depend on the distance travelled divided by the inelastic mean free path, $\exp\left((-z/\cos\theta)/\lambda\right)$. Thus di is given by

$$di = A\,C_X J \sigma_{x,y} \frac{\Omega}{4\pi} G_{(E)} \exp\left(\frac{(-z/\cos\theta)}{\lambda}\right) p\,r\,dz \qquad (3.64)$$

where A is the irradiation area, J is the incident electron (or possibly X-ray) flux, $\sigma_{x,y}$ is the electron impact cross-section (which takes into account how easy it is to get Auger electron emission from a given set of levels), p is the probability of initially ionizing the ion core (usually $\sim$1) and r is the backscattering factor which takes into account additional ionizations by secondary electrons.

The expression given in eqn (3.64) is then integrated across all z, since it has the form

$$\int di = K C_x \int_0^\infty \exp\left(\frac{(-z/\cos\theta)}{\lambda}\right) dz \tag{3.65}$$

where K sums all the other terms, which are constants for a given spectrometer. Integration yields the expression

$$i_A = C_x \lambda \frac{\Omega}{4\pi} A G_{(E)} \cos\theta J \sigma_{x,y} p r \tag{3.66}$$

C_x can be found from the Auger current, as all the other parameters can be measured. The expression for i_{PES} is given in eqn (3.67) and differs from that for the Auger current in the last four terms in eqn (3.66). In PES, J is the incident X-ray flux and $\sigma_{x,y}$ is the photoionization cross-section which is different from that in the Auger process. r is irrelevant in the case of PES as no further photoelectrons are produced from secondary processes. p is taken to be 1 for PES.

$$i_{PES} = C_x \lambda \frac{\Omega}{4\pi} A G_{(E)} \cos\theta J \sigma_{x,y} \tag{3.67}$$

The use of AES and PES for quantitative analysis is difficult because a considerable number of approximations often have to be made to determine C_x. This is best done by using ratios of currents obtained from samples of known concentration against measurements of the sample under consideration, recorded by the same spectrometer, using exactly the same conditions. It should be noted that the escape depth or inelastic mean-free-path of the emitted electrons is much the same as those evolved in PES (as described in Section 2.2.4) and so AES spectra are found to be of similar surface sensitivity.

3.6.1.3 *Chemical shift*

The AES electrons are affected by the chemical environment of the element; this can be thought of as slightly altering the energy of the levels E_x, E_y, and E_z by a small equal amount, ΔE:

$$E_{Auger} = E_x - E_y - E_z \rightarrow (E_x + \Delta E) - (E_y + \Delta E) - (E_z + \Delta E) \tag{3.68}$$

so that the total shift in Auger energy should be ΔE.

This 'chemical shift' from the expected elemental position is reflected in both the position and shape of the observed peak. In this way AES spectra are analogous to PES spectra. However, the situation for Auger transitions is not a simple as in PES because in real AES spectra the energy levels redistribute themselves to compensate for the loss of the electrons and so this expression is more of an approximation.

It turns out that the shape of a given Auger peak is of importance because it is highly dependent on the structure of the adsorbate and so can be used as a 'fingerprint' of the adsorbate molecules. For example, the shape of the peaks due to adsorbed CO are vastly different from those of adsorbed ethene. Figure 3.45 shows the shapes of the AES peaks

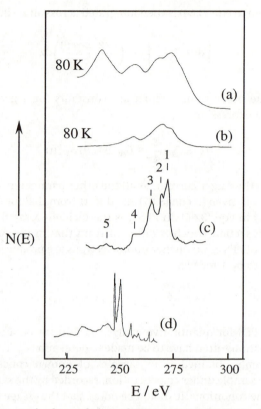

Fig. 3.45 AES spectra in the form of plots of $N(E)$ against E for C adsorbed in three different chemical environments at 80 K: (a) ethene/Pt(111) (b) ethane/Pt(111) and (c) carbon monoxide/Cu(111). (d) The AES spectrum of gas-phase CO. ((a) and (b) from Canning *et al.* 1981, (c) from Baker *et al.* 1981, and (d) from Moddenham *et al.* 1971.)

for the carbon atom in three different adsorbed molecules: ethene and ethane on Pt(111), carbon monoxide on Cu(111) and gas-phase CO. The spectra are plotted in the form $N(E)$ against E, to show the different peak shapes which result for the carbon atoms in these very different chemical environments. (The spectra were first obtained in the form $dN(E)/dE$ against E and the sharp features were then reintegrated by computer so that the peaks do not appear on a sloping background.)

3.6.1.4 *Scanning Auger microscopy (SAM)*

It is possible to use AES to obtain a chemical 'map' of the surface. The incident electron beam can be focused down to a very small spot size (about 1 μm in diameter) and then scanned across the surface using a rastering power supply like that used in a television. (In a television set, the raster scanner moves the electron beam rapidly across the phosphor dots on the screen to excite them. The beam is usually scanned in horizontal lines from left to right and top to bottom across the screen.) By using a computer to

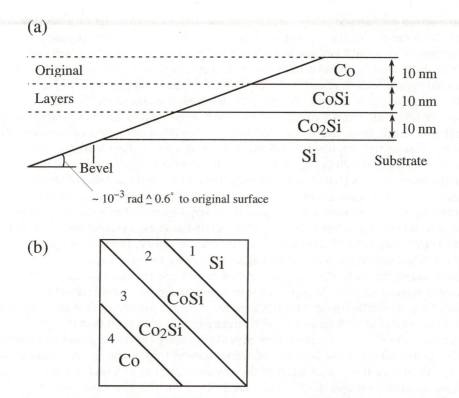

Fig. 3.46 (a) Schematic of the structure of the Co/Co₂Si/CoSi/Si layers, into which a bevel has been cut at an angle of ~0.5° to the surface. (b) Plan view of the expected chemical composition.

correlate the position of the beam with the Auger electrons evolved, it is then possible to build up a map of the chemical composition of the surface. Usually several scans are taken, with the analyser set to a different Auger peak energy for each scan; thus for scanning an Si peak, at regions of high Si concentration, a bright patch is observed in the image.

The example which is given in this section is typical of the sort of interfaces which are used in the electronics industry. Electronic devices are produced by growing conducting layers on semiconductor wafers and etching away part of the layers to leave conducting wires and pads on the surface (in a process called metallization). As many as seven or eight layers can often be deposited to form a single integrated circuit. It is very difficult to study the interfacial regions between the layers at the points at which they have been etched because most techniques have insufficient spatial resolution. However, this sort of information is vital to our understanding of how the device works and for determining how to improve the performance of a device by producing 'sharper' interfaces. SAM measurements can be of considerable value in this respect.

In the example, SAMs were taken of a layered structure formed from 10 nm of Co deposited on 10 nm of Co_2Si, deposited on 10 nm of CoSi deposited on an Si substrate as shown in Fig. 3.46(a). The SAM images recorded are shown in Fig. 3.47. Figure 3.47(a) is an electron microscope image recorded by the Auger spectrometer of a bevel which has been cut out of the layered structure. The bevel was cut very precisely with a computer-controlled 2 keV xenon ion beam and produced an area of exposed surface at an angle of roughly 1 mrad (0.6° to the original surface) as indicated in Fig. 3.46(a). This should expose regions of Co, Co_2Si, CoSi and Si as shown in plan view in Fig. 3.46(b). The image in Fig. 3.47(a) was recorded by detecting the 100-eV electrons leaving the sample (these electrons have an energy which is not due to an Auger transition in either Si or Co). From the image it is clear that the bevel has produced a crater with steep sides on the top, bottom and left side, while a lower slope is seen on the right side. The images shown in Fig. 3.47(b)–(d) are for an area along the bottom of the image of the crater in (a); each uses the scale indicated on the right-hand side of the figure to show the relative amounts of the element present. Red indicates large amounts (high intensity) and blue, small amounts (low intensity). Figure 3.47(b) is the image formed by detecting the Co AES peak of the MVV transition at 45 eV, (c) shows the intensity for the Si LVV peak at 85 eV and (d) the Co LMM peak at 766 eV. The low-energy region of the AES spectra (in the form of $N(E)$ against E, where the peaks appear on a sloping background) for each of the regions 1 to 4, which are indicated in Fig. 3.47(b), are shown in Figs 3.47(e)–(h). In region 1 the only peak observed is for Si, while in region 4, only Co is seen. The two spectra recorded for points at the centre of regions 2 and 3 have peaks due to both Co and Si, with larger amounts of Si in (g) than (f). It turns out that from the intensities of the Co and Si peaks (measured by taking peak heights once the spectra have had the background slopes removed) in (f) and (g), in combination with the images, the stoichiometry of the exposed layers can be found. Region 3 should have a composition of Co_2Si, and this is found to be the case from the AES measurements. Region 2 should be due to a CoSi area, but it is found that the composition varies rapidly across the region from very Co 'rich' in the area next to region 3 to almost no Co close to region 1.

Another development of SAM is to use a rapidly scanning AES system to obtain the most important peaks, and then to sputter through the sample and rapidly scan again, revealing a depth profile of the elemental concentrations as a function of depth.

3.6.2 Secondary ion mass spectrometry (SIMS)

If a beam of ions such as O^+ or Ga^+ is fired at a surface at an energy of several keV, the surface is sputtered away as described in Section 2.5.2. If a representative fraction of the sputtered species is detected by a mass spectrometer, it is possible to piece together the chemical composition of the surface and any layers of adsorbates. The method is called secondary ion mass spectrometry (SIMS) and has very high sensitivity to chemicals on the surface, far higher than AES. It is, by its nature, a highly destructive technique and the detector response to different elements may vary by several orders of magnitude. On the other hand, the advantage of using this technique is that there is an enormous amount of information contained in the spectra. For example, if you were investigating the adsorption of a polymer on a surface, it is possible to detect all the fragments that have been sputtered off and, from these, piece the structure together.

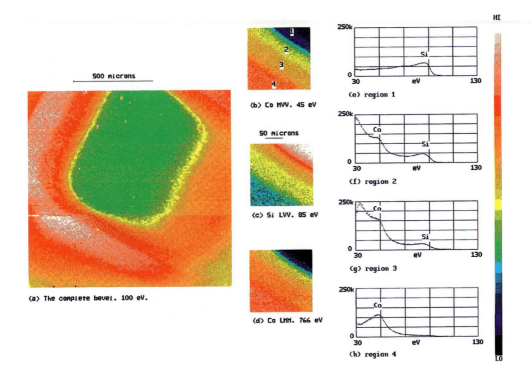

Fig. 3.47 Energy analysed and SAM images of the Co/Co₂Si/CoSi/Si layer structure shown in Fig. 3.46. (a) An image recorded at 100 eV, showing the whole of a bevel which has been etched out of the structure. (b)–(d) SAMs of a region along the bottom of the bevel in (a), recorded for AES peaks at (b) Co at 45 eV, (c) Si at 85 eV, and (d) Co at 766 eV. (e)–(h) The AES spectra of the central areas of regions 1–4 which are marked on (b). The right of the figure shows the rainbow colour scale which is used to indicate the relative intensity of the peak being recorded. Red shows the highest intensity (or amount) and blue the lowest intensity (or amount). (From Prutton 1994).

There are two types of SIMS:

(i) Static SIMS, in which only the top layer is removed for analysis, and

(ii) Dynamic SIMS, in which several layers are removed in a manner similar to that for depth profiling AES. The beam is scanned across the surface in the same way as the electron beam is scanned across the phosphor on the TV screen and the fragments correlated in three dimensions. (It should be noted that the rastering of the beam is not in the same pattern as used for a TV. The patterns used for SIMS are more complex to avoid allowing the surface to rearrange itself during the scan and hence give distorted information. Segregation of components in the surface and differential sputtering are two serious problems in dynamic SIMS.)

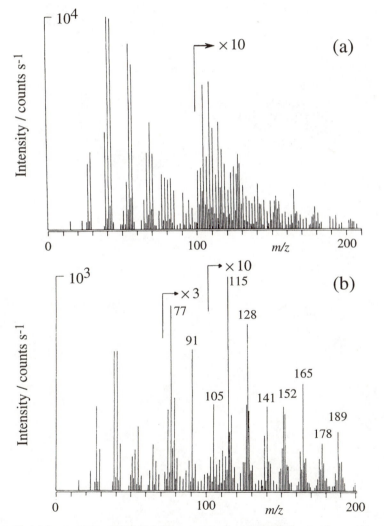

Fig. 3.48 (a) Static SIMS of PVC. (b) The same surface as (a) following beam damage by 2 keV Xe^+ ions. (From Briggs and Hearn 1986.)

Dynamic SIMS is used as one of the major methods of quality control in the semiconductor industry, where some of the wafers being produced on the production line are measured for purity and accuracy of levels of doping.

An example of a static SIMS result is shown in Fig. 3.48(a) for the surface of a polyvinyl chloride (PVC) sample. Figure 3.48(b) shows the effect of beam damage on the surface by a beam of xenon ions. The large number and mass distribution of the fragments evolved show how complex SIMS spectra can be and indicate the vast amount of information which is contained in them.

4

Surface reactions and reactivity

4.1 Reactions

Following adsorption, as we have seen in Section 3.5 it is possible for an adsorbate to undergo a rearrangement, to fragment on the surface or to simply be desorbed in its original molecular form. It is also possible for the adsorbate to react with another species which is either co-adsorbed on the surface or in the gas phase. As we have seen in Chapter 1, it is this reactivity which is makes surfaces and reactions on or by them of such great importance in everyday life. In fact, one of the major motivations for studying surface reactions has been to apply our knowledge to the field of catalysis. A useful working definition of a catalyst is *a substance which increases the rate of a chemical reaction without being consumed* and much of the work described will be set in the context of catalysis.

In the chapter we will explore the mechanisms of surface reactions, discussing single molecular species and reactions involving more than one. The product formed and the nature of its adsorption is also of importance because this can limit the use of certain catalysts. For instance, it may bind very effectively to the surface and prove difficult to remove. Studies of catalysts tend to explore their macroscopic properties. For example, how efficient is the catalyst? How stable? How robust? How durable? In surface science studies we concentrate on the microscopic aspects of the heterogeneous reaction. For example, how does the catalyst work? What is the adsorption process and what is the nature of the adsorbate on the surface? From the previous chapter it is clear that in order to answer some of these important questions, the same techniques used to study adsorption can also be applied successfully to the study of surface reactions and to the detection of reaction intermediates.

One difficulty in relating surface science measurements to those in catalysis is that real catalysts work under considerably higher pressure conditions than it is possible to measure with surface science tools. For example, a typical catalysed process operates at well over an atmosphere in pressure while typical surface science experiments can only be carried out in vacua better than 10^{-6} mbar. This means that there is a pressure difference of 9–10 orders of magnitude between the two, which may well result in different behaviour. This gap is beginning to close, however, as technology begins to allow us to use high-pressure cells for adsorption and to develop techniques that use photons as the probe species. These act in a complementary manner to charged particles but do not require high vacua. In fact, IR techniques have been applied very successfully to observe the same adsorbate species *in situ* on a catalyst and on single-crystal surfaces. Kinetic data can be obtained for both surface science and 'real' catalysis experiments and can be analysed in exactly the same way to provide valuable insights into reaction mechanisms.

It is also important to realize that studies of surface reactions are highly important in fields other than catalysis. In particular, two of the most important are the semiconductor industry and in environmental chemistry where the study of heterogeneous processes in the atmosphere has proved enormously valuable.

4.2 Transition state theory

For a reaction in the gas phase or on the surface where

$$A + B \rightarrow C \tag{4.1}$$

the rate can be found from the general expression

$$\text{Rate} = k_n C_A^x C_B^y \tag{4.2}$$

where C denotes the concentrations of the components A and B. x and y are the experimentally determined orders of reaction in A and B, respectively. k_n is the rate coefficient which is related to an activation energy E^* and pre-exponential factor A by:

$$k_n = A \exp\left(\frac{-E^*}{RT}\right) \tag{4.3}$$

If the reaction is a single-step one, the rate would be proportional to $C_A C_B$, that is with the values of $x = y = 1$. However, the reaction normally proceeds in a series of elementary steps, in which case the observed rate equation depends on all the steps up to the slowest rate-determining one. Thus, for surface reactions the exponents can be non-integral, or even negative when one adsorbate is strongly adsorbed and excludes another.

When comparing the homogeneous process in the gas phase to the heterogeneous process involving the surface, it is found that the rate is increased for the heterogeneous process, either because A is larger or, more commonly, because the surface enables E^* to be smaller or even zero. E^* is usually reduced because the surface provides a different pathway by which the reaction can proceed. It should be noted that the surface has no effect on the position of the equilibrium, which depends entirely on the enthalpy and entropy of the overall reaction.

Figure 4.1 shows the potential energy curve illustrating the energetics involved when using transition state theory to describe how the reaction proceeds. In the homogeneous process, the reactants have a substantial activation energy barrier E_{hom}. However, once they reach the transition state by overcoming this, the final product is formed rapidly.

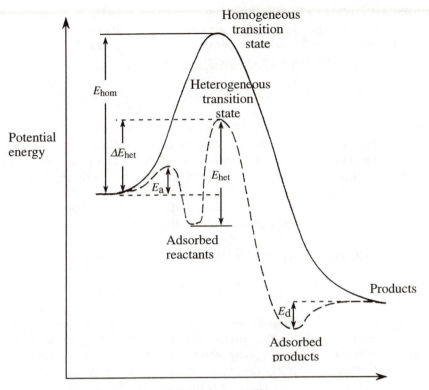

Fig. 4.1 A potential energy diagram, using transition state theory to compare the homogeneous and heterogeneous reaction pathways. (Solid lines indicate the homogeneous process and dashed lines show the heterogeneous process.)

For the comparable heterogeneous process, the reactants initially overcome the small activation energy barrier for adsorption E_a, and are adsorbed on the surface. In order for the reaction to proceed, the adsorbed reactants then have to overcome the activation energy barrier E_{het} to form the heterogeneous transition state, which normally requires considerably less energy than formation of the homogeneous transition state (i.e. $E_{het} < E_{hom}$). The products are then formed as adsorbates on the surface and they desorb once they have overcome E_d, the activation energy barrier for desorption of the product species. The total energy difference between the initial reactants and final products in the gas phase is the same for both homogeneous and heterogeneous processes, which is why the equilibrium position is the same for each. Making the assumption that the rate-determining step is excitation to the transition state, the reaction proceeds faster on the surface because of the relative sizes of E_{het} and E_{hom}. It should be noted from eqn (4.3) that in cases where the rate of a surface reaction is temperature dependent, the rate depends on both the value of E_{het} and the temperature T.

It is not too surprising to note that the overall reaction rate for the surface reaction is dependent on the surface area of the catalyst. This means that in order for it to be worthwhile to carry out a heterogeneous reaction rather than the homogeneous one, there must be both sufficient surface available and a sufficient reduction in the activation energy barrier. To illustrate this we will look at the bimolecular reaction in eqn (4.1). The rate of the homogeneous reaction is given by

$$R_{\text{homo}} = C_A C_B A \exp\left(\frac{-E_{\text{hom}}}{RT}\right) = C_A C_B \frac{kT}{h} \frac{[q_{\neq}]_{\text{hom}}}{q_A\, q_B} \exp\left(\frac{-E_{\text{hom}}}{RT}\right) \qquad (4.4)$$

where C_A and C_B refer to the gas-phase concentrations of adsorbates A and B per unit volume. The pre-exponential factor, A, is determined from the partition functions, q, which sum the contributions of translational, electronic, vibrational and rotational states in statistical thermodynamics; q_A and q_B are the partition functions of components A and B and $[q_{\neq}]_{\text{hom}}$ is the partition function for the homogeneous transition state, where one degree of freedom (motion along the reaction coordinate) is factored out. (k is the Boltzmann constant, T the temperature and h the Planck constant.)

For the heterogeneous reaction, a similar expression is obtained:

$$R_{\text{hetero}} = C'_A C'_B C'_S \frac{kT}{h} \frac{[q_{\neq}]_{\text{het}}}{q_A\, q_B} \exp\left(\frac{-E_{\text{het}}}{RT}\right) \qquad (4.5)$$

C'_A and C'_B refer to concentrations per unit area of adsorbed species, but for our purposes we will assume that the surface species are in concentrations which are directly proportional to the gas-phase concentration, thus $C_A = C'_A$ and $C_B = C'_B$. Making the further assumption that reaction takes place only between adjacent adsorbed A and B molecules, the term C_S, the concentration of adjacent pairs of surface sites, must be introduced. This is true if both A and B are adsorbed on the surface (the so-called Langmuir–Hinshelwood reaction mechanism, see Section 4.5.1). $[q_{\neq}]_{\text{het}}$ is the partition function for the heterogeneous transition state. In order to assess the difference in activation energies required to make the heterogeneous process viable, we take the ratio of the two rates:

$$\frac{R_{\text{hetero}}}{R_{\text{homo}}} = C_s \frac{[q_{\neq}]_{\text{het}}}{[q_{\neq}]_{\text{hom}}} \exp\left(\frac{\Delta E}{RT}\right) \qquad (4.6)$$

where

$$\Delta E = E_{\text{hom}} - E_{\text{het}} \qquad (4.7)$$

$[q_{\neq}]_{\text{hom}}$ includes translational motion whereas $[q_{\neq}]_{\text{het}}$ does not, because translational motion is lost on the surface. Since translational energies are considerably larger than vibrational, rotational, *etc.*, the following substitution can be made:

$$\frac{[q_{\neq}]_{\text{het}}}{[q_{\neq}]_{\text{hom}}} \approx \frac{1}{q_{\text{trans}}} \approx \frac{1}{7 \times 10^{26}} \qquad (4.8)$$

where q_{trans} is the partition function for the translations. A value of $\sim 7 \times 10^{26}$ is typical for a fairly light gas molecule such as D_2 at room temperature (for our purposes, this was calculated for a molecule confined in a 100 cm^3 vessel). The value of q_{trans} indicates that 7×10^{26} quantum states are accessible. Substitution for the ratios of the partition functions given in eqn (4.8) into eqn (4.6) then gives

$$\frac{R_{\text{hetero}}}{R_{\text{homo}}} = C_{\text{s}} \frac{1}{q_{\text{trans}}} \exp\left(\frac{\Delta E}{RT}\right) \tag{4.9}$$

As we have seen in Chapter 2, a surface has approximately 10^{14}–10^{15} sites cm^{-2}. Taking 5×10^{14} as a typical value of C_{S}, we find that to obtain equal rates, $\Delta E = 69.8$ kJ mol^{-1} for a total surface area of 1 cm^2, at 300 K. In other words, under these conditions the activation energy barrier for the heterogeneous process must be at least 69.8 kJ mol^{-1} lower than that of the homogeneous process for the heterogeneous process to be viable. However, it should be noted that considerably smaller reductions than this are usually sufficient because real working catalysts have very large surface areas (often hundreds or thousands of square centimetres).

4.3 Selectivity

In the previous section we saw that a heterogeneous process can accelerate a reaction. Another important feature of surface reactions is that they can provide a degree of selectivity in the products, by making accessible different possible reaction pathways. Reactions can be sequential, that is of the form

$$A \rightarrow B \rightarrow C \tag{4.10}$$

or parallel, where different products are produced on the same surface.

$$\begin{aligned} A &\rightarrow B \\ &\rightarrow C \end{aligned} \tag{4.11}$$

In the former case, catalysts can be chosen to select the reaction product B rather than C or vice versa. An example of a sequential reaction is the oxidation of hydrocarbons such as ethene. The reaction is given in eqn (4.12) and a schematic potential energy curve and the most significant states formed are given in Fig. 4.2.

$$C_2H_4 + \frac{1}{2}O_2 \rightarrow C_2H_4O \xrightarrow{+O_2} H_2O + CO_2 \tag{4.12}$$

For the real process there are number of intermediate steps, but the key role of the catalyst is in reducing the activation energy barrier of the first step $(C_2H_4 + \frac{1}{2}O_2 \rightarrow C_2H_4O)$ to almost zero. The catalyst has no effect on the second major step in the reaction, that is from the C_2H_4O intermediate to the final states. Water and carbon dioxide are the most thermodynamically stable products, but the catalysed reaction can be used to selectively enhance the less stable products by speeding up their formation rate relative to their destruction rate.

The role of the substrate material (or materials, in the case of real catalysts) is vital in determining the products formed. A prime example of this is in the reactions of

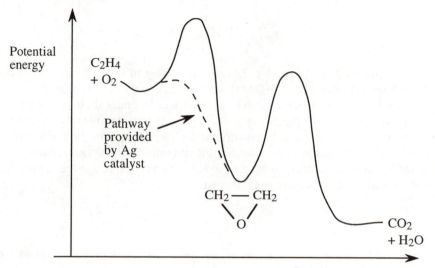

Fig. 4.2 A schematic potential energy curve for the oxidation of ethene. (Solid lines indicate the homogeneous process and dashed lines show the heterogeneous process.)

synthesis gas, which is a mixture of carbon monoxide and hydrogen. It is usually made by passing steam over coal. Figure 4.3 shows the products resulting from passing synthesis gas over several different metals.

The reaction of CO and hydrogen on copper produces methanol. This reactivity results from the nature of the interaction of CO with copper. The d orbitals of copper are full and so the interaction with CO is weak; too weak, in fact, to enable the $C\equiv O$ bond to be broken. The steps which are likely to be involved in the process are shown in Fig. 4.4. Initially it is thought that CO and hydrogen adsorb onto the copper surface. H then binds sequentially to the carbon atoms, first forming adsorbed HCO, then CH_2O, and finally methoxide, CH_3O. In the final step of the process, an OH species is formed and the methanol is then readily desorbed.

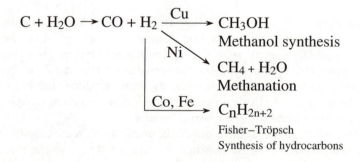

Fig. 4.3 The reaction scheme for the synthesis gas reaction over different metals.

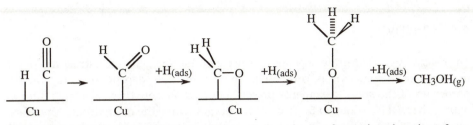

Fig. 4.4 Schematic diagram of the generally accepted mechanism for methanol synthesis from synthesis gas over a copper-based catalyst.

Methanol synthesis over copper-based catalysts is of such great industrial importance that a massive amount of research has been conducted on it over the past twenty or more years, using a variety of surface science techniques. It should be noted that in the real catalytic process, the gaseous mixture which is passed over the supported copper catalyst contains not only carbon monoxide and hydrogen but also carbon dioxide. The experimental evidence suggests that the mechanism that occurs over the real catalyst involves hydrogenation of the carbon dioxide to formate, followed by stepwise hydrogenation to methanol. This process is more energetically favourable than the process involving only CO and hydrogen.

The transition metals Ni, Co and Fe are more reactive than Cu because their partially filled d orbitals are able to take part in bonding to the adsorbates. These partially filled d-bands enable the metals to break the C≡O bond in carbon monoxide. (The C≡O bond is one of the strongest bonds known; the bond energy is $\sim 10^3$ kJ mol^{-1}.) Under appropriate conditions, particularly in terms of temperature, each of these metals is found to completely dissociate the CO, forming adsorbed C and O on the surface. (For example, nickel can only dissociate CO above room temperature.)

The generally accepted mechanism for methanation by a nickel-based catalyst is shown in Fig. 4.5. Following the initial dissociative chemisorption, adsorbed hydrogen atoms bind sequentially with the carbon atom until a CH_3 adsorbate species is formed. Binding to a further hydrogen atom results in the formation of methane which is readily desorbed to the gas phase. The Fischer–Tröpsch synthesis of hydrocarbons over Co- or Fe-based catalysts is found to proceed similarly. Polymerization of the CH_n species to form heavier hydrocarbon fragments on the surface is found to be favoured on these surfaces.

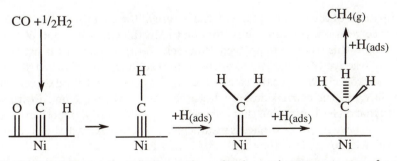

Fig. 4.5 The generally accepted reaction pathway for the methanation reaction of synthesis gas by a nickel-based catalyst.

4.4 Reactivity

The reactivity of a substrate depends on the electronic and geometric structures of the materials from which it is composed. These structures depend not only on the properties of the specific elements present but also on how these elements interact with each other (see Chapter 5). Two key factors governing the reactivity are, therefore, the nature of the elements in the substrate and the crystal faces which are exposed. Volcano curves can be used to compare the reactivities of different transition metals and are discussed in Section 4.4.1. In Chapter 3 we saw that the structure of a substrate can have a considerable effect on adsorption; in Section 4.4.2 the effect of substrate structure on the reactivity will be discussed.

4.4.1 Volcano curves

From the surface reactions described in Section 4.3, it is apparent that there is a link between the nature of the chemisorption of a species on the surface and its reactivity. If chemisorption is 'weak' then low coverages of adsorbate will be formed and the surface reactivity (or catalytic activity) will be low. As the heat of adsorption increases, bonding to the surface becomes stronger and the uptake increases, giving higher surface coverages and reduced energies of activation. Thus, the catalytic activity will be greater. At the other extreme, if an adsorbate interacts too strongly with the surface, the latter will quickly become covered by adsorbed species and the surface–adsorbate system may then be too stable to decompose or react further. The catalytic activity will, in this case, be low.

The catalytic activity of a surface for a given reaction can be plotted against the atomic number (or a similar suitable parameter such as the d-band occupancy) to show how the activity changes with substrate across the periodic table. Looking at the example of the Fischer–Tröpsch synthesis of hydrocarbons which was given in Section 4.3, Fig. 4.6 shows a schematic plot of the catalytic activity against atomic number. It is clear that the catalytic activity is low for Sc on the left-hand side of the periodic table. Activity increases to a maximum at Co and then decreases again towards copper on the right-hand side of the period. This type of plot is typically known as a 'Balandin volcano plot'.

A comparison of the volcano plot with the nature of the chemisorption of CO shows a distinct correlation. For substrates from Sc to Mn the chemisorption of CO is too strong to allow this reaction to proceed. However, adsorption on Fe, Co and Ni gives a sufficiently reactive CO adsorbate to enable it to react once chemisorbed. In terms of the Fischer–Tröpsch synthesis the catalytic activity is greater on Fe and Co compared to that on Ni. While the conversion of hydrocarbons does take place on Ni with similar activity to that on Fe, it turns out that the preferred product on Ni is methane itself. On copper, as we have seen, the surface is insufficiently reactive to crack the CO and so the catalytic activity for hydrocarbon synthesis is very low.

Similar volcano plots are found for most catalytic processes. Obviously they peak at different metals depending on the process under investigation. For example, for

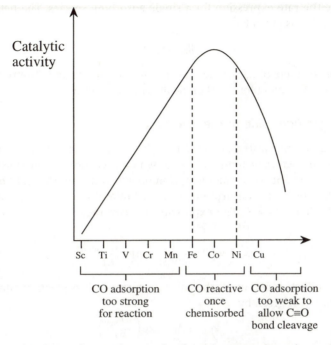

Fig. 4.6 A typical 'volcano plot'; in this case for the Fisher-Tröpsch synthesis of hydrocarbons from synthesis gas, with a correlation to the bonding of CO on the metals.

methane synthesis from synthesis gas, the volcano plot is strongly peaked at Ni and for methanol synthesis the volcano plot is peaked at copper. These volcano curves can therefore be used to select the most suitable metal for carrying out a desired surface reaction. Generally they correlate the strength of chemisorption with the reactivity. The peak usually occurs at some region of intermediate strength of chemisorption, with reactivity being low at the extremes where very weak and very strong adsorption is found.

The relative catalytic activities for different metals can be characterized by the temperature at which a given rate of reaction is attained (lower temperatures representing higher activities) or by the rate coefficient (k) for the reaction at a given temperature (a high rate coefficient representing high activity). The use of k is not altogether satisfactory however. From eqn (4.3) we know that k involves both A and E^*, and these vary with substrate. If both vary together (i.e. both vary in the same direction) on changing substrate, they counteract each other (the so-called 'compensation effect', where the change in A compensates for the change in E^*). This is, in practice, a common finding and may give a misleading view of the relative catalytic activities of the surfaces.

A more satisfactory method of quantifying the volcano plot is to combine the Langmuir isotherm (given in eqn (3.33) in Section 3.3.1) with transition state theory (Section 4.2) to investigate the rates of reaction across the periodic table. The starting point is to use the rate expression for a single adsorbing species. The heterogeneous reaction rate R_{hetero} is given by

$$R_{\text{hetero}} = k_{\text{het}}\,\theta \tag{4.13}$$

where θ is the coverage of adsorbate and k_{het} the rate coefficient. The right- and left-hand sides of the volcano plot are then dealt with separately.

4.4.1.1 The right-hand side of the volcano plot

To look at reactions on substrates on the right-hand side of the volcano curve, use can be made of the characteristic temperatures at which a reaction on one substrate equals the rate on another. For weak chemisorption on substrates on the right-hand side of the volcano plot, the heat of adsorption, ΔH_{ads}, is low and so b (see eqn (3.32) in Section 3.3.1) is small. Thus $bP << 1$. The expression for coverage θ can then be written in terms of P and b:

$$\theta = bP = \frac{P}{\text{const}}\exp\left(\frac{-\Delta H_{\text{ads}}}{RT}\right) \tag{4.14}$$

where 'const' collects together all the constant terms in b as given in eqn (3.32). The rate coefficient k_{het} is given by

$$k_{\text{het}} = A_{\text{het}}\exp\left(\frac{-E_{\text{het}}}{RT}\right) \tag{4.15}$$

Feeding eqns (4.14) and (4.15) into eqn (4.13) results in

$$R_{\text{het}} = A_{\text{het}}\frac{P}{\text{const}}\exp\left(\frac{-\Delta H_{\text{ads}} - E_{\text{het}}}{RT}\right) \tag{4.16}$$

To compare two substrates X and Y, we can take ratios of expression (4.16), as it applies to the two substrates. This gives the two temperatures T_X and T_Y, at which their rates R_X and R_Y are equalized.

Because the exponential terms normally dominate, the pre-exponential factors can be ignored. When $R_X = R_Y$, the two exponential terms are equal and so

$$\frac{\Delta H_X + E_{\text{hetX}}}{T_X} = \frac{\Delta H_Y + E_{\text{hetY}}}{T_Y} \tag{4.17}$$

where the heat of adsorption ΔH and activation energy E_{het} for X and Y are indicated by subscripts. What we want to know is how the heat of adsorption plus activation energy barrier for substrate X varies compared to that on substrate Y. An example of this would be on going from a copper catalyst (X) to a nickel one (Y) when considering the synthesis gas reactions. (CO is more strongly bound on nickel.) Rearranging eqn (4.17) and for convenience making the assumption that $E_{\text{hetX}} \approx E_{\text{hetY}}$ enable us to simplify the expression to eqn (4.18). Note that this is not a very likely situation but adopting it allows us to draw some reasonable comparisons between substrates.

$$\Delta H_X + E_{\text{hetX}} = \frac{\delta\Delta H_{\text{ads}}\,T_X}{\Delta T} \tag{4.18}$$

where $\Delta T = T_Y - T_X$ and $\delta\Delta H_{\text{ads}} = \Delta H_Y - \Delta H_X$.

Equation (4.18) can thus be used to compare adjacent substrates on the right-hand side of the volcano curves. For simple single-step processes, the left side of eqn (4.18) is always positive because although heats of adsorption are negative, the activation energy barrier is always positive and larger than the heat of adsorption. For the right side of eqn (4.18), an increase in adsorbate interaction with the surface on going from X to Y (or Cu to Ni in our example) gives a negative value of $\delta\Delta H_{ads}$ which means that ΔT must also be negative. Thus $T_Y < T_X$ or, in our example, $T_{Ni} < T_{Cu}$, which *is* observed. In addition, the linear relationship between $\delta\Delta H_{ads}$ and ΔT is apparent and can be seen in the slope on the right-hand side of the volcano curves.

It should be noted that in complex multistep processes, the overall activation energy barrier may be smaller than the heat of adsorption because of changes in energy affecting the bottom of the potential energy curve.

4.4.1.2 *The left-hand side of the volcano plot*

On the left-hand side of the volcano plot, ΔH_{ads} is large because chemisorption is strong. This means that $bP >> 1$ and so $\theta \approx 1$. Thus, eqn (4.13) reduces to

$$R_{hetero} = k_{het} \tag{4.19}$$

and the reaction is therefore of zero order. Considering eqn (4.15), this implies that the activation energy for the reaction E_{het} is the only energy term which affects the rate and that changing ΔH_{ads} does not. Obviously this is not what is observed and we must look more closely at E_{het}.

In this case, it is thought that changes in the energy with which the molecule binds to the surface affect the bottom of the potential energy curve rather more than they affect the transition state. It is thus assumed that the heterogeneous activation energy, E_{het}, is dependent on both the heat of adsorption and the energy difference between the gas-phase reactants and the transition state, ΔE_{het}. ΔE_{het} is shown in Fig. 4.1 and it is this energy which is thought to be insensitive to the strength of the chemisorption. E_{het} is given by

$$E_{het} = -\Delta H_{ads} + \Delta E_{het} \tag{4.20}$$

Thus the rate can be given by

$$R_{hetero} = k_{het} = A_{het} \exp\left(\frac{-\Delta H_{ads} + \Delta E_{het}}{RT}\right) \tag{4.21}$$

It should be noted that for the right-hand side of eqn (4.21) there is a roughly linear relationship between ΔH_{ads} and the temperature. Bearing this in mind, if we consider the implications of eqn (4.21) it is apparent that as the heat of adsorption becomes less negative (i.e. strength of adsorption decreases), the total activation energy E_{het} decreases, so the reaction goes faster. This expression therefore satisfactorily models the behaviour observed on the left-hand side of the volcano curve, in the region of strong chemisorption.

4.4.1.3 *Two-reactant systems*

For a situation in which there are two adsorbates A and B on the surface, as in eqn (4.1), the rate expression is expected to take a general form

$$R = k\,\theta_A\,\theta_B \tag{4.22}$$

Volcano plots can be produced for the reaction of A and B. The strengths and relative strengths of chemisorption of both components dictate how efficient the reaction process will be. If one of the components is too strongly adsorbed in comparison to the other, for example if A is more much more strongly bound than B, then A will quickly produce a coverage of $\theta_A \approx 1$. There will consequently be no sites available for B to adsorb and the reaction will not proceed. If it does proceed, with difficulty, the rate will depend on θ_A^{-1}.

4.4.2 Structure sensitivity

One area in which surface science studies have usefully been applied to heterogeneous catalytic processes is in the investigation of reaction rate as a function of the surface structure. Studies of the reaction rate have been carried out on a range of flat (low-indexed), stepped and kinked surfaces. Typical reactions studied include aromatization (e.g. *n*-hexane to benzene and *n*-heptane to methyl benzene) and cyclization (e.g. ethyne to benzene). It is found that these reactions proceed much faster on the (111) faces of fcc metals such as platinum compared to their rates on (100) faces. For the aromatization reactions it has been found that the hexagonal (111) face is between three and seven times more active than the square-based (100) mesh. This activity difference is due to the type and nature of the adsorption site(s) available for both the reactants and the products formed. A hexagonal substrate mesh is particularly suitable for reactions where benzene is formed because the substrate offers a 'template' for binding of the appropriate atomic dimensions and geometry. In other words, benzene 'fits' neatly into the hexagonal structure of the Pt(111) surface.

Further work on aromatization on Pt has shown that even faster rates are obtained for reactions on stepped and kinked surfaces, with maximum rates being achieved on stepped surfaces with hexagonal terraces about five atoms wide. The role of steps and kinks is not yet fully understood but it is clear that they often provide regions of considerable reactivity. The difference in their chemical behaviour compared to that of the terraces is due to their very different local atomic structural environments which cause them to have very different electronic charge densities.

The surface structure (both electronic and geometric) can be modified by adsorption of another species to alter the reactivity or to improve efficiency of a given reaction. For example, if potassium is adsorbed on a metal such as Fe, it donates electrons to the iron which increases the negative charge available for bonding to adsorbates. This is borne out by the observation that CO is more strongly bound on the K-modified surface (its CO stretching frequency is lower) than on the clean surface. K is often used as an electronic promoter, with its most important use, as K_2O, on the Fe-based ammonia synthesis catalyst.

Structure modifiers are also widely used, usually to block sites which are too reactive. For example, when carrying out the hydrogenolysis of organic molecules on

Pt, kink and step sites often produce undesirable low molecular weight products. It is found that H_2S reacts on Pt to form adsorbed S. The S is much more strongly bound to the steps and kinks than to the terrace sites. Controlled exposure to H_2S can therefore be used to block these sites. Such selective poisoning of the kinks and steps leaves the terraced sites available to produce the desired higher molecular weight products.

4.5 Reaction mechanisms

It is important that we have some idea of the mechanisms by which surface reactions proceed and much of the early insight into these came from the study of their kinetics. There are two generally accepted models for surface reactions, namely the Langmuir–Hinshelwood (LH) and the Rideal–Eley (RE) mechanisms. Observations of how changes in pressure and temperature affect the reaction rate can be used to distinguish between the two mechanisms and these are described in detail in Sections 4.5.1 and 4.5.2. It should be noted that these simple kinetic studies can be ambiguous to some degree (see Sections 4.5.3–4.6). Rather more sophisticated models and methods are usually needed to properly elucidate a given surface reaction mechanism. These limitations are discussed further in Section 4.5.3.

4.5.1 The Langmuir–Hinshelwood (LH) mechanism

In this reaction scheme all reactants involved in the reaction are adsorbed on the surface prior to reacting. When using simple kinetics to study this mechanism, several assumptions are made:

 (i) the reaction at the surface is the rate-determining step of the overall reaction;
 (ii) the Langmuir isotherm can be applied to all gases involved in the reaction (and so its assumptions also apply—see Chapter 3, Section 3.3.1);
(iii) adsorbates compete for the same surface sites.

The mechanism can be discussed for a number of different situations.

4.5.1.1 Unimolecular surface reaction with immediate desorption of the product

In this case a single species, A, is adsorbed and either undergoes a rearrangement or decomposition to form a product B which is immediately desorbed, that is

$$A_{(gas)} \rightarrow A_{ads} \rightarrow B_{(gas)} \tag{4.23}$$

Substitution for θ_A from the Langmuir isotherm in Section 3.3.1, eqn (3.33), gives an expression for the rate of reaction R:

$$R = k_{het}\,\theta_A = \frac{k_{het}b_A P_A}{1 + b_A P_A} \tag{4.24}$$

There are two possible limiting cases:

(i) where b_A (and thus ΔH_{ads}) is small (weak chemisorption) or P_A is small, so that $b_A P_A \ll 1$. The rate in expression (4.24) reduces to

$$R = k_{het}\, b_A\, P_A \qquad (4.25)$$

which is first order in pressure of A;

(ii) where b_A (and thus ΔH_{ads}) is large (strong chemisorption) or P_A is large, so that $b_A P_A \gg 1$. The rate in expression (4.24) then reduces to

$$R = k_{het} \qquad (4.26)$$

so the rate is zero order in pressure of A. Under these conditions the coverage of A on the surface will be close to a monolayer, independent of the pressure of A.

4.5.1.2 Molecular rearrangement/decomposition—product also adsorbed

In this case the reactant A is adsorbed, undergoes a rearrangement or decomposition to form product B and this is also adsorbed in competition with A, before it finally desorbs:

$$A_{(gas)} \rightarrow A_{ads} \rightarrow B_{ads} \rightarrow B_{(gas)} \qquad (4.27)$$

The rate of reaction here depends on the competition between A and B adsorbed on the surface. Substituting for the coverage of A from the Langmuir expression for this situation, which was given in Chapter 3, eqn (3.35), leads to the rate expression

$$R = k_{het}\, \theta_A = \frac{k_{het} b_A P_A}{1 + b_A P_A + b_B P_B} \qquad (4.28)$$

There are two limiting cases:

(i) At low pressures of A, where P_A is low, $b_A P_A \ll 1$ and so the rate reduces to

$$R = \frac{k_{het} b_A P_A}{1 + b_B P_B} \qquad (4.29)$$

(ii) When b_B is large, the product is much more strongly bound to the surface than the initial reactant A. In this case, $b_B P_B \gg b_A P_A \gg 1$ and the rate is

$$R = \frac{k_{het} b_A P_A}{b_B P_B} \qquad (4.30)$$

$k_{het}\, b_A/b_B$ is known as the composite rate coefficient in this case, and the product B 'poisons' the surface.

4.5.1.3 Bimolecular surface reactions

For the reaction scheme involving two reactants on the surface,

$$A_{(gas)} + B_{ads} \rightarrow A_{ads} + B_{ads} \rightarrow C_{ads} \rightarrow C_{(gas)} \qquad (4.31)$$

applying the Langmuir isotherm gives an overall expression for the rate R as

$$R = k_{het}\,\theta_A\,\theta_B = \frac{k_{het}b_A P_A b_B P_B}{(1 + b_A P_A + b_B P_B + b_C P_C)^2}$$

(4.32)

where b_C and P_C are the Langmuir b factor and pressure of the product C, respectively. The appearance of these parameters arises because the total coverage of A is affected by the amount of C on the surface, thus

$$\theta_A = \frac{b_A P_A}{1 + b_A P_A + b_B P_B + b_C P_C}$$

(4.33)

The expression for the coverage of B is analogous. Using eqn (4.33) it can be seen that again there are several limiting cases.

(i) When the product is either desorbed immediately or is weakly bound with respect to the reactants so that $b_A P_A \approx b_B P_B \gg b_C P_C$, then the reaction rate is found to pass through a maximum at $b_A P_A = b_B P_B$.

(ii) When the product is either desorbed immediately or is weakly bound with respect to the reactants, and the reactants are both weakly bound such that $b_A P_A \ll 1$ and $b_B P_B \ll 1$, then the rate becomes

$$R = k_{het}\,b_A\,P_A\,b_B\,P_B$$

(4.34)

that is it is second order in gas pressure.

(iii) When the product is either desorbed immediately or is weakly bound with respect to the reactants and one reactant (A) is much more weakly bound than the other (B), then $b_A P_A \ll b_B P_B + 1$. The rate becomes

$$R = \frac{k_{het}\,b_A P_A\, b_B P_B}{(1 + b_B P_B)^2}$$

(4.35)

Extending this to the situation where B is very strongly bound so that $b_B P_B \gg 1$ gives

$$R = \frac{k_{het}\,b_A P_A}{b_B P_B}$$

(4.36)

In this case the reaction is first order in the pressure of A but inhibited by increasing the pressure of B. Because B is far more strongly bound it inhibits the reaction because it blocks the adsorption sites, preventing A from adsorbing.

(iv) If the product is strongly adsorbed with respect to the reactants so that $b_C P_C \gg 1 + b_A P_A + b_B P_B$ and $b_C P_C \gg 1$ then the rate becomes

$$R = \frac{k_{het}\,b_A P_A\, b_B P_B}{b_C^2 P_C^2}$$

(4.37)

So the reaction is strongly inhibited by the product which is said to 'poison' the reaction, again by blocking sites for adsorption.

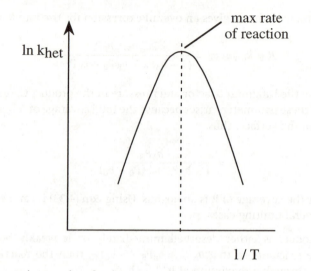

Fig. 4.7 The Arrhenius plot of ln k_{het} versus $1/T$ for a typical surface reaction.

4.5.1.4 *Maximum rate surface reactions as a function of temperature*

Arrhenius plots of ln k against $1/T$ are often produced for gas-phase reactions where generally the reaction rate is found to increase with temperature. An Arrhenius plot can also be made for a surface reaction, but it is often found to pass through a maximum, as shown schematically in Fig. 4.7. The reason for this observation is that the overall rate achieved is a balance between the tendency for the surface reaction rate to increase with temperature (with characteristic activation energy E_{het}) and for the surface coverage to decrease with increasing temperature.

We can consider the example in Section 4.5.1.3(iii) above where A is weakly bound and B is strongly adsorbed on the surface. The overall rate is given in eqn (4.35). At high pressure and low temperature (which models the right side of the graph in Fig. 4.7) $b_B P_B \gg 1$ and the rate is given in eqn (4.36). k_{het} depends exponentially on $-E_{het}$ (eqn 4.15) and b depends exponentially on ΔH_{ads} (eqn 3.32). The slope of the plot is determined by the observed activation energy E_{tot} where

$$E_{tot} = E_{het} + \Delta H_{ads-A} - \Delta H_{ads-B} \tag{4.38}$$

B is more strongly adsorbed then A and so $\Delta H_{ads-A} - \Delta H_{ads-B} > 0$ because heats of adsorption are negative and $\Delta H_{ads-A} \ll \Delta H_{ads-B}$. This means that E_{tot} is positive and so the Arrhenius plot has the expected slope where k_{het} increases with temperature, or ln k_{het} decreases with $1/T$, just as is seen for the negative slope on the right side of the graph in Fig. 4.7.

For the left side of the graph in Fig. 4.7 the conditions are such that $b_B P_B \ll 1$ (for low pressures and high temperatures) and eqn (4.35) reduces to eqn (4.34). The slope here is determined from E_{tot} which in this case is

$$E_{tot} = E_{het} + \Delta H_{ads-A} + \Delta H_{ads-B} \equiv E_{het} - q_{ads-A} - q_{ads-B} \tag{4.39}$$

It is possible that $-\Delta H_{ads-B} > E_{het}$ and so by this point E_{tot} could have become negative. Under these conditions the reaction goes slower at higher temperatures; in other words ln k_{het} increases with $1/T$. This model adequately describes the positive slope on the left side of the plot in Fig. 4.7.

4.5.2 The Rideal–Eley (RE) mechanism

The Rideal–Eley mechanism was postulated to take into account the observation that not all the reactants taking part in the surface reaction need to be adsorbed first. It is possible in some reactions for a gas-phase molecule B to collide with an adsorbed molecule A as in

$$A_{ads} + B_{(gas)} \rightarrow C_{ads} \rightarrow C_{(gas)} \tag{4.40}$$

The rate R' is then given by

$$R' = k'_{het}\, \theta_A\, P_B \tag{4.41}$$

where k'_{het} is the heterogeneous rate coefficient for the Rideal–Eley mechanism. Substituting for θ_A (which was defined in eqn (3.33)) and making the assumption that the product C is either immediately desorbed or only weakly adsorbed in comparison to the strength of adsorption of A gives

$$R' = \frac{k'_{het}\, b_A P_A\, P_B}{1 + b_A P_A} \tag{4.42}$$

This means that the overall rate is always first order in pressure of the gas-phase component B. There are two limiting cases for A:

(i) where b_A (and thus ΔH_{ads}) is small (weak chemisorption) or P_A is small, so that $b_A P_A \ll 1$. The rate expression becomes

$$R' = k'_{het} b_A\, P_A\, P_B \tag{4.43}$$

so the reaction is first order in pressure of A;

(ii) where b_A (and thus ΔH_{ads}) is large (strong chemisorption) or P_A is large, so that $b_A P_A \gg 1$. The rate in expression becomes

$$R' = k'_{het}\, P_B \tag{4.44}$$

so the reaction is zero order in pressure of A.

For the Rideal–Eley mechanism, just as for the Langmuir–Hinshelwood mechanism, the kinetics are further complicated if the product C is strongly bound and poisons the reaction. In that case the expression for θ_A is

$$\theta_A = \frac{b_A\, P_A}{1 + b_A\, P_A + b_C P_C} \tag{4.45}$$

and this can be fed into eqn (4.41) to give

$$R' = \frac{k'_{het}\, b_A\, P_A\, P_B}{1 + b_A\, P_A + b_C P_C} \tag{4.46}$$

Similar limiting cases can then be found.

In principle, kinetics only gives information up to and including the slowest step. From the above it can be seen that from careful studies of the pressure and temperature dependence of the rate of a surface reaction, information relevant to the mechanism (be it Langmuir-Hinshelwood or Rideal–Eley) can be deduced. In addition, it is possible to use the kinetic data to obtain the numerical values of the activation energy for the reaction, E_{het}.

4.5.3 Limitations of simple kinetic methods

In Sections 4.5.1 and 4.5.2 the two simplest models for reaction mechanisms were described. While the simple kinetics approach can often be very useful to give initial insights into mechanisms, real reaction mechanisms are often far more complex and this often leads to significant problems associated with fitting kinetic data to simple models. These problems arise in part because all the assumptions of the Langmuir isotherm are required to hold, that is the surface must be homogeneous, the heat of adsorption on a given site must be independent of the surrounding adsorbates and each site is occupied by only one adsorbate. In addition, in the case of fast surface reactions it is likely that the ad-layer is not in equilibrium with the gas phase, which is implicit for assessing the Langmuir-Hinshelwood mechanism when using a kinetic approach. However, these simple kinetic models provide a framework which includes all the important relevant parameters for deducing a quantitative relationship to describe the observed kinetics. In that way they are analogous to the continued use of the Langmuir isotherm for adsorption studies.

It is also clear that many kinetic studies in the literature have used these simple models and it is important to have a firm understanding of how the data have been interpreted. It is probably fair to say that the simple kinetic information described in Sections 4.5.1 and 4.5.2 *might* be rather misleading when used to determine reaction mechanisms, particularly in heterogeneous catalytic processes. For example, the kinetics might indicate that a Langmuir-Hinshelwood mechanism is in operation, whereas the real mechanism might be Rideal–Eley (see Section 4.5.5). In Sections 4.5.4, 4.5.5 and 4.6, examples are given which show how other methods are needed to reveal the complexity of, and gain clear insights into, surface reactions.

4.5.4 Temperature programmed reaction spectroscopy (TPRS)

Temperature programmed reaction spectroscopy (TPRS) is a fairly sophisticated mass spectrometric tool for investigating reactions on surfaces. The method is very like TPD, which was described in Section 3.4.2, in that the adsorbate-covered surface is heated rapidly and the desorbing molecules are detected. The difference is that in TPRS a number of mass peaks (often as many as 16) are monitored, so that any product(s) and reactants evolved can be determined.

TPRS can therefore be used to give information on the selectivity of a given substrate, as data can be obtained following adsorption under a range of different conditions. In addition, each component of a complex surface system can be studied separately to try to determine the role each plays in the reaction.

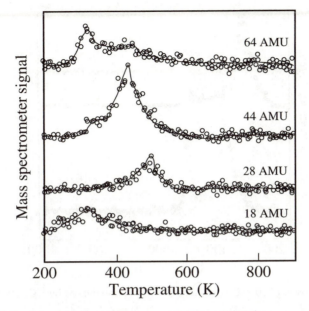

Fig. 4.8 TPRS following adsorption of propane at 160 K on Pt(111), precovered by O_2 and SO_2 dosed at 300 K. (From Wilson *et al*. 1995.)

These two key features of the method are clearly illustrated by an investigation to gain an understanding of the roles of promoters and poisons on catalysts. In the first example, TPRS was applied to investigate the reactivity on a typical automobile exhaust catalyst; specifically to probe the effect of sulphur dioxide on the oxidation of propane over platinum. Propane does not adsorb on a Pt(111) surface which has been pre-covered with either oxygen or sulphur dioxide; the TPRS recorded for the oxygen-covered surface contain no peaks due to CH combustion (propane does not dissociate readily, because of the stability of the CH bond). Likewise, TPRS for the SO_2-covered surface contains only a single peak which is due to desorbing SO_2 at 64 amu. Figure 4.8 shows the TPR data recorded at 18 (H_2O), 28 (CO), 44 (CO_2) and 64 (SO_2) amu following exposure of a Pt(111) surface to O_2/SO_2 at 300 K and adsorption of propane at 160 K. It is clear that there is a small amount of SO_2 desorption at 300 K and that H_2O, CO_2 and CO desorb at 320, 430 and 500 K, respectively. The O_2/SO_2 precoverage was therefore found to 'activate' the surface towards the dissociative chemisorption of propane; once adsorbed it was oxidized by the O-containing adsorbed species. By varying the initial adsorption conditions it was found that the surface was only activated by O_2/SO_2 preadsorption in the temperature range 250–400 K. On the basis of supporting experiments using XPES and EELS, the 'activating' species was postulated to be SO_4.

Further information on the mechanisms of the reactions can be gained from TPRS by the use of isotope tracing experiments. In the second example of TPRS, shown in Fig. 4.9, the oxidation of CO on the O_2/SO_2 pre-dosed Pt(111) surface was studied by adsorbing $^{18}O_2$ with $S^{16}O_2$ on the surface at 300 K, prior to exposing it to CO at 160 K. TPR data were recorded at 44 ($C^{16}O_2$) and 46 ($C^{16}O^{18}O$) amu as shown. The appearance

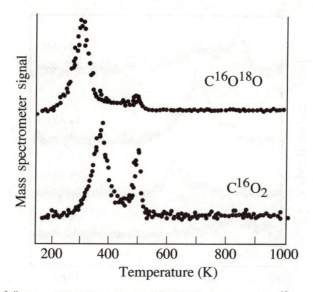

Fig. 4.9 TPRS following 160 K CO exposure of Pt(111) precovered by $^{18}O_2$ and $S^{16}O_2$. (From Wilson *et al.* 1996.)

of peaks for both these species show that there are two routes by which the CO_2 is formed on the surface. The route by which $C^{16}O^{18}O$ is formed must involve adsorbed ^{18}O, while the route to $C^{16}O_2$ must involve oxygen from the $S^{16}O_2$.

4.5.5 Oscillatory reactions: CO oxidation on Pt/Pd

A reaction of considerable importance both in catalysis and from a fundamental point of view is that of the oxidation of CO. The reaction proceeds very efficiently over the surfaces of the noble metals palladium and platinum, with a rate proportional to P_{O_2}/P_{CO} (the partial pressures of oxygen and carbon monoxide, respectively). Remarkably, this result was found for Pd and Pt catalysts at the beginning of the twentieth century.

4.5.5.1 *Mechanism of CO oxidation on Pt/Pd*

In 1969, early surface science work by Ertl's group used a mass spectrometer to monitor the rate of carbon dioxide production on Pd(110) as a function of O_2 and CO partial pressures and substrate temperatures. LEED measurements were made simultaneously to monitor the structure on the surface. In this very forward-looking study the LEED patterns were correlated with LEED patterns observed after separate adsorption of the reactants. A typical plot for CO_2 production as a function of temperature at constant partial pressures of the reactants is shown in Fig. 4.10. Below 100°C the LEED was typical of a CO covered surface, indicating that high coverages were present; the adsorption of oxygen was apparently inhibited by the presence of the CO.

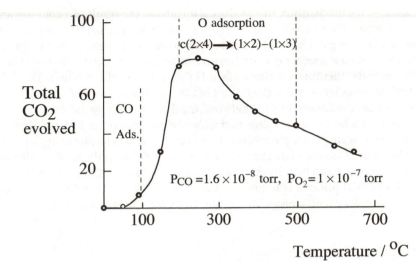

Fig. 4.10 The variation of the rate of carbon dioxide production from a Pd(110) surface as a function of substrate temperature. The CO and O_2 partial pressures were held constant at 1.6×10^{-6} and 1×10^7 torr, respectively. (From Ertl and Rau 1969.)

As the substrate temperature was increased, some of the CO was desorbed, the resulting stationary CO coverage was reduced and the inhibiting effect for oxygen adsorption was decreased. The reaction requires chemisorbed O atoms on the surface and so the reaction rate increased continuously with oxygen coverage until a maximum was reached between 200 and 300°C. The rate decreased again above this temperature as the oxygen coverage decreased, this time because of oxygen desorption. In this temperature regime the coverage of CO was also very low. Again LEED data proved valuable for determining the structures formed on the surface (this time predominantly due to O atoms) and thus the coverages of reactants. Descriptions of the LEED patterns observed are also indicated in Fig. 4.10.

Measurements of the variation of the rate of production of CO_2 as a function of partial pressure of O_2 and CO at constant temperature confirmed the previous observations that the rate of the reaction r is proportional to P_{O_2}/P_{CO}. On this basis it was suggested that a Rideal–Eley mechanism was in operation, with O atoms adsorbed and CO reacting from the gas phase.

From more detailed work it then became clear that the reaction was much more complex than this. It was found that CO forms densely packed ad-layers which did indeed inhibit the dissociative chemisorption of oxygen. Chemisorbed O atoms were found to form open structures which did not have a noticeable affect on the uptake of CO. In view of these observations, it turns out that if simple kinetic models are applied, it is impossible to distinguish between the two mechanisms. Even in the case of the Langmuir–Hinshelwood mechanism, the rate would not be expected to decrease with increasing partial pressure of oxygen. It therefore became apparent that these stationary measurements were inadequate and that some time-resolved method was needed.

In the Rideal–Eley mechanism, the product is formed as the result of a collision and this takes place on a timescale of $\sim 10^{-12}$ s; this is much shorter than the flight time of the molecules in the gas phase and can be viewed as being instantaneous. In the Langmuir–Hinshelwood mechanism, both reactants are adsorbed first and should have measurable average lifetimes on the surface before forming the product. Thus the detection of a finite delay between the impact of CO with the surface and the evolution of CO_2 proves that the Langmuir–Hinshelwood mechanism is in operation.

Time-resolved measurements using molecular beams are described in detail in Section 4.6.3. In this case a simple measurement can be carried out by using a pulsed molecular beam of CO to correlate the arrival time of the pulse on the surface with the desorption time of the product. In the case of the oxidation of CO on Pd and Pt, the residence time of CO prior to reaction was found to be 5×10^{-4} s, which effectively rules out a Rideal–Eley mechanism. This is not, however, the end of the story!

4.5.5.2 *Oscillatory behaviour*

The oxidation of CO on Pd and Pt is even more complex than the above description might suggest. In the early 1970s, the rate of reaction under continuous flow conditions was found to oscillate. This is shown in Fig. 4.11 for the reaction over a polycrystalline Pt wire. In Fig. 4.11 the partial pressures of the reactants were found to oscillate 180° out-of-phase with the rate of formation of carbon dioxide. The work function change was found to parallel the reaction rate and could be correlated with the changes in oxygen coverage on the surface. When work was carried out to investigate the reaction on Pt(100), CO-induced structural changes were also found to coincide with the oscillations.

In a remarkable study using photoemission electron microscopy (PEEM), Ertl's group followed the oxidation on Pt(110) in real time and discovered the extraordinary effects shown in Fig. 4.12. PEEM can be used on a mezoscopic length scale and is a contrast technique based on the measurement of differences in the local work function. This means that areas of the surface at different stages of adsorption/reaction can easily be distinguished. In Fig. 4.12, four images are shown, recorded after 30 s intervals. Areas of the surface which are oxygen covered emit few electrons and so appear dark on the fluorescent screen, whereas areas predominantly covered by CO emit more electrons and so appear light. The spiral patterns formed expand with time and are currently the subject of considerable analysis and theoretical modelling.

4.6 Dynamics of surface reactions

The kinetics of a surface process describe the influence and effect of the macroscopic properties involved in a reaction, for example pressure and temperature and their effect on the reaction rate. The kinetics of surface reactions was one of the first features

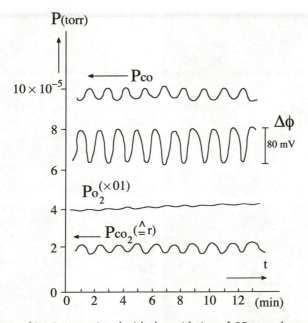

Fig. 4.11 Oscillatory kinetics associated with the oxidation of CO on polycrystalline Pt wire under steady-state flow conditions. The substrate was held at a temperature of 502 K and the incident partial pressures of CO and O_2 were 1×10^{-4} and 4×10^{-4} torr, respectively. (From Ertl *et al.* 1982.)

studied as it was the most accessible experimentally; conventional gas-phase kinetics techniques involving the pressures of reagents and products could be applied readily.

Of increasing interest and importance within our knowledge of surfaces and their behaviour is an understanding of the detail of the atomic motions of the surface and adsorbates during a reaction. These are the *dynamics* of a surface process and the concept was introduced in Section 3.1.5. Studies of surface dynamics, especially of surface reaction dynamics, are becoming increasingly prominent as techniques to study them become more and more feasible. Surface reactions take place on a large variety of timescales, often in the range from as much as 10^3 s down to about 10^{-12} s and for the latter, ultrafast techniques are necessary. The developments in and applications of molecular beam and laser techniques to surface systems have proved particularly valuable for such purposes.

At our current state of knowledge, much of our understanding of the energetics of a given reaction comes from the measurement of the surface kinetics followed by the fitting the data to simple postulated rate laws as described in Section 4.5. As we have seen, these studies can be misleading and it is vital to investigate the dynamics of surface processes, which provide the key to understanding them.

Dynamics studies are particularly advantageous in determining the details of the surface–adsorbate and adsorbate–adsorbate interactions. However, such experiments are usually demanding to perform and often require a large amount of theoretical input for proper interpretation. Dynamics encompasses the whole range of surface

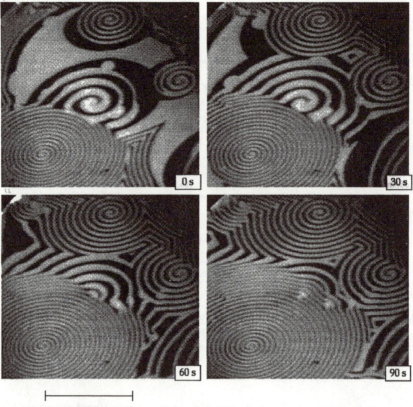

Fig. 4.12 PEEM images from a Pt(110) surface, recorded during the catalytic oxidation of CO at a substrate temperature of 448 K. The CO and O_2 partial pressures were held constant. The images exhibit the evolution of oxygen-concentration spirals with varying wavelengths and rotation periods. (From Nettesheim *et al.* 1993.)

processes, including adsorption, desorption, diffusion and reaction. Consequently, it requires an understanding of the behaviour of the ground and excited translational, electronic, vibrational and rotational states involved. The pathways and rates of energy transfer, particularly of vibrational energy flow between the surface and adsorbate, are of major importance because this type of energy transfer often provides the driving force behind the surface processes.

The situation for chemical reactions is far more complex than for the simple molecular dissociation process described in Section 3.1.5. Many degrees of freedom are often involved in the reaction pathway and other degrees of freedom are required to describe the pathways available to remove any excess energy. Many of these possible pathways are vibrational in character and can involve the internal vibrations of the adsorbate, the vibrations of the surface itself (the phonons) and the frustrated motions (translations and rotations) of the adsorbate bonded to the surface.

4.6.1 Laser methods

The energy flow between all of these modes controls how thermal equilibrium is maintained during a reaction. If there is a large deviation from equilibrium during the reaction, it is often found that the flow of vibrational energy controls the rate of reaction. Dramatic deviations from equilibrium can be achieved by using rapid heating with nanosecond, picosecond or even femtosecond laser pulses. The effect of fast heating on the resulting surface reaction is often spectacular. It is found that, depending on the laser frequency used, some degrees of freedom obtain higher effective 'temperatures' than others (that is become populated at higher energies). This means that these types of experiment can provide excellent insights into the competition between dissociation and desorption pathways for a diatomic molecule; providing additional energy to certain degrees of freedom may promote one pathway rather than another. For example, supplying energy to the surface–adsorbate bond promotes desorption, while additional energy in the molecular vibrational modes promotes dissociation. With sufficiently fast lasers it *might* be possible, in the future, to achieve bond-selective dissociation.

4.6.2 Atomic/molecular beams

Molecular beams provide a different method of following reaction dynamics. The energy transfer processes occurring at surfaces which have most influence on the adsorption/desorption or reaction pathways occur under non-equilibrium conditions, when high temperatures and pressures of the adsorbing gas hit the relatively cold surface. In order to model such conditions, a molecular or atomic beam is fired at a surface and the scattered molecules or atoms are spatially and energetically analysed. This is state-selective spectroscopy where the final analysis provides information on the rotational, vibrational, translational and, sometimes, electronic energies of the scattered molecules. In a typical experiment a molecular beam is produced by expanding tens of atmospheres of gas through a small ($\sim 30\ \mu$m) nozzle, as shown in Fig. 4.13. The beam undergoes 'adiabatic' cooling during expansion, such that most of the energy available is channelled into translational energy in the direction of the beam. When the resulting speed of the beam is greater than that of sound, the beam is described as being supersonic; such beams have the special property of being almost monoenergetic. This arises because there are such a large number of molecules in a confined space that the collisional mean free path is very short and so the molecules undergo many collisions per unit time. They rapidly reach a local thermodynamic equilibrium within the beam, with the result that all molecules have very similar energies. Although they have large imposed translational energies, the molecular beams have low vibrational and rotational energies (or effective 'temperatures' since the populations of the rotational and vibrational energy levels can be described in terms of the temperature at which such Boltzmann populations would be normal).

The beam is pulsed by passing it through a rotating chopper. The starting time of the pulse is measured. Following scattering from the surface, the pulse is detected, usually by a time-of-flight (TOF) mass spectrometer, and the arrival time is recorded. The difference between start and arrival times often includes some 'dwell' time, which

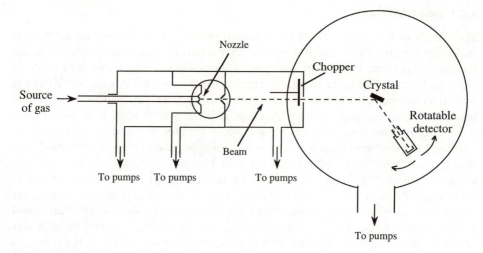

Fig. 4.13 A typical atomic/molecular beam scattering apparatus.

results (for example, see Section 4.5.5.1) from the scattering molecules interacting with the surface. The dwell time is usually zero if the gas–surface interaction occurs as a single inelastic scattering event. (It should be noted that elastic scattering of supersonic He beams results in the detection of the diffraction pattern of the surface as described in Section 2.6.2.2.) If the axes of the mass spectrometer can be rotated, the final velocity distribution can be obtained as a function of the scattering angle.

It is also possible to determine the final rotational and vibrational state distribution of the molecules in the beam by using, for example, laser-excited fluorescence (LEF) or multiphoton ionization (MPI). For these methods, a tunable laser beam is passed through the scattered beam. The chosen laser frequency only excites one set of the vibronic levels (J,v) to higher levels (these are the vibrational + rotational quantum levels which are denoted by the quantum numbers v for vibrational and J for rotational levels). Once excited, they decay either by fluorescence (the fluorescence emission is monitored by LEF) or by ionization by a second laser photon (the ion flux is monitored by MPI). The signal detected in either case is proportional to the initial population of the original (J,v) levels which were excited. If the tunable laser is swept across its frequency range then excitation from different vibrational + rotational levels can take place and so a complete energy map of the distribution of the final molecular states in the beam can be obtained. The distributions obtained are characteristic of the types of interaction the reactive beam has undergone with the surface, how much energy exchange has occurred with the surface and by what mechanism. Possible scattering mechanisms include energy exchange with single or multiple phonons of the metal surface (so-called direct-inelastic scattering) or via a sticking–trapping channel.

4.6.3 HD formation on Cu(111) by the Rideal–Eley mechanism

For reactions, molecular beam methods can be applied to give insights into both the mechanism and the energetics involved. In an elegant study, Rettner and Auerbach investigated the dynamics of formation of HD from a beam of D(H) atoms colliding with H(D) adsorbed on the Cu(111) surface. The results show that the reaction proceeds via the Rideal–Eley (RE) mechanism. The HD product was detected by a directional TOF mass spectrometer to obtain angular distributions and by MPI to obtain the quantum level distributions of the HD molecules. Fig. 4.14(a) shows the angular distributions for a beam of D atoms incident on an H-covered surface at 100 K, for two different incident beam energies $E_i = 0.3$ and 0.07 eV. The distributions of product molecules are asymmetric about the surface normal, showing that they have a 'memory' of the direction of the incident beam which implies that this is an RE mechanism. If it were a Langmuir–Hinshelwood (LH) mechanism no such memory is expected as both adsorbates must bond to the surface before they can react. It is also found that the distributions are different for H beams on D-covered surfaces, which supports the analysis that this reaction has an RE mechanism; if it were an LH mechanism the distributions would be the same.

The velocity distribution of the HD showed that the product was formed with a wide range of energies, independent of the angle at which it was scattered. The quantum state distributions (i.e. the different vibrational + rotational levels occupied) are shown in Fig. 4.14(b) for HD molecules produced from D atoms colliding with an H-covered surface, and in Fig. 4.14(c) from H atoms colliding with a D-covered surface. The plots take the form of the flux of molecules with energies in levels J and v, $F(v, J)$; the internal energy is found by summing the vibrational and rotational energies. The average vibrational and rotational energies are found to be different depending on whether the reaction was carried out using D on an H-covered surface or H on a D-covered surface. For D on H/Cu the average rotational energy is lower than for H on D/Cu while the average vibrational quantum number is higher. The sum of the rotational and vibrational energies with the observed translational energy of the HD is found to be equal to the sum of the heat of adsorption of H or D with the incident translational energy of the D or H beams. This shows that the reaction can couple efficiently to all three of the product energy channels, rotation, vibration and translation. The energies are of relative sizes $E_{trans} > E_{vib} > E_{rot}$ as expected.

From careful and complex analysis of all of the data, significant specific information about the reaction can be obtained. The information on the degree of vibrational excitation can be used to assess the form of the potential energy surface for the reaction. The data on the relative occupation of vibrational and rotational levels can be used to assist in modifying theoretical models.

4.7 Heterogeneous reactions in the atmosphere

In the past two decades the chemistry taking place in the atmosphere has received increasing attention, largely motivated by our fears about environmental pollution. Initial work concentrated on reactions in the gas phase, as gas-phase species constitute

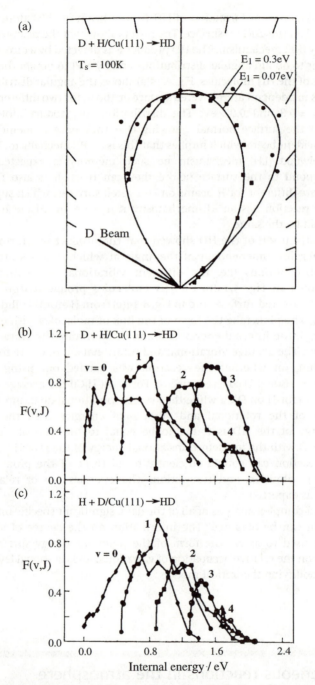

Fig. 4.14 The reaction of atomic beams of H and D with the D- and H-covered Cu(111) surface. (a) The TOF distribution of HD scattered from the surface when a D beam was scattered off an H-covered surface, (b) the quantum state distributions (in terms of J and v) of the HD product of the interaction of D atoms with an H-covered Cu(111) surface, and (c) as for (b) but for H atoms on a D covered surface. (From Rettner and Auerbach 1996.)

the majority of the components of the atmosphere. However, attention has since focused on heterogeneous reactions after it became clear that gas-phase models alone could not account for some of the observations being made.

It has been found that heterogeneous reactions (involving gas–solid, liquid–solid and gas–liquid interfaces) are important in both the lower atmosphere where we live (the troposphere) and in the upper atmosphere (the stratosphere). In both regions heterogeneous chemistry has far-reaching effects, but the reactions taking place are quite different. In the troposphere heterogeneous reactions are important in the formation of acid rain and in the production of pollutants in smog, etc. Their effects, although devastating, are usually fairly localized. Of more global importance is the effect of heterogeneous reactions in the stratosphere which, for example, are known to cause the depletion of the ozone layer.

4.7.1 The ozone hole

In order to appreciate the importance of this research area, it is useful to give a brief introduction to stratospheric ozone chemistry. The ozone (O_3) 'layer' exists in the stratosphere, between altitudes of 15 and 40 km. Rather than being a real layer, the amount of ozone present in fact constitutes only a few parts per million (ppm) of the stratospheric gases, but it is of vital importance to life on earth. This is because stratospheric ozone acts as a filter, stopping radiation at high energies in the UV/C and B regions from reaching the surface of the earth. UV/C and B radiation cause severe damage to organisms because they degrade DNA; it is thought that life only developed on the earth's surface once sufficient ozone became present in the stratosphere to offer protection. (It should be noted that molecular oxygen O_2 also filters out some of the UV/C.) Ozone works as a filter because of its photochemistry. It exists in a dynamic equilibrium, where it is created and destroyed, in the following cycle, which was first devised by Chapman in 1930.

$$
\begin{aligned}
O_2 + h\nu\,(< 246\text{nm}) &\rightarrow O^\bullet + O^\bullet \\
O^\bullet + O^\bullet + M &\rightarrow O_2 + M \\
O^\bullet + O_2 + M &\rightarrow O_3 + M \\
O^\bullet + O_3 &\rightarrow O_2 + O_2 \\
O_3 + h\nu\,(< 290\text{nm}) &\rightarrow O_2 + O^\bullet
\end{aligned}
\tag{4.47}
$$

where M is a third body. In the 1970s it was realized that the Chapman cycle did not account for the total amounts of ozone present in the stratosphere. The levels were measured and found to be much lower than expected. It was realized that small concentrations of other chemically active species in the stratosphere, at $\sim$10–100 times lower concentrations than ozone, were involved in cycles which effectively reduced the ozone present at equilibrium. The most important of these cycles involve NO_x, ClO_x and HO_x species. The NO_x and ClO_x cycles are the two on which most attention has been focused and they proceed as follows.

$$
\begin{aligned}
\text{NO}_x \quad & \text{NO} + \text{O}_3 \rightarrow \text{NO}_2 + \text{O}_2 \\
& \text{NO}_2 + \text{O}^\bullet \rightarrow \text{NO} + \text{O}_2 \\
\text{Overall}: \quad & \text{O}_3 + \text{O}^\bullet \rightarrow 2\text{O}_2
\end{aligned}
$$

$$
\begin{aligned}
\text{ClO}_x \quad & \text{Cl}^\bullet + \text{O}_3 \rightarrow \text{ClO}^\bullet + \text{O}_2 \\
& \text{ClO}^\bullet + \text{O}^\bullet \rightarrow \text{Cl}^\bullet + \text{O}_2 \\
\text{Overall}: \quad & \text{O}_3 + \text{O}^\bullet \rightarrow 2\text{O}_2
\end{aligned}
\tag{4.48}
$$

The amount of ozone in the stratosphere is reduced to a fairly constant value through the Chapman cycle in conjunction with cycles such as those in eqns (4.48). It was noted that the latter cycles 'neutralize' each other to a certain extent because they interact with each other to form 'reservoir' species such as $ClONO_2$, N_2O_5 and HCl. Once the reservoir species are formed they do not interfere with the ozone chemistry.

In 1979 the British Antarctic Survey team began to observe that depletion of ozone in the stratospheric region above the South Pole was beginning to occur just after the Antarctic winter; the depletion has continued to occur annually since then. By the middle of the 1980s it became clear that gas-phase models for atmospheric chemistry could not explain this depletion and it was suggested that heterogeneous processes were involved. The formation of the ozone hole is now generally thought to occur through the following complex process:

(i) Vortex formation. This is shown in Fig. 4.15 and arises because in the polar winter the air over the pole becomes cold and dark. It is largely isolated from the rest of the atmosphere for the long winter period.

(ii) Polar stratospheric cloud (PSC) formation. The low temperatures in this part of the stratosphere lead to condensation of water to form PSCs and thus the atmosphere is dehumidified.

(iii) Denitrification. NO_x species are removed from the atmosphere by adsorption on the PSCs to form HNO_3 on the surface. NO_x species are therefore no longer available to neutralize the ClO_x species by forming the reservoir species.

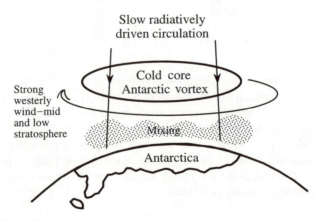

Fig. 4.15 Vortex formation over the Antarctic.

(iv) Preconditioning. The PSCs also provide a surface capable of returning inactive chlorine contained within the existing reservoir species back into active chlorine species, which can then cause ozone destruction once sunlight returns in the spring. There are thought to be four main adsorbate reactions implicated here, as follows:

$$HCl_{(ads)} + ClONO_{2(g)} \rightarrow Cl_{2(gas)} + HNO_{3(ads)} \qquad (4.49)$$
$$H_2O_{(s)} + ClONO_{2(g)} \rightarrow HOCl(g) + HNO_{3(ads)} \qquad (4.50)$$
$$HCl_{(ads)} + N_2O_{5(g)} \rightarrow ClNO_{2(g)} + HNO_{3(ads)} \qquad (4.51)$$
$$H_2O_{(s)} + N_2O_{5(g)} \rightarrow 2HNO_{3(ads)} \qquad (4.52)$$

(v) Spring. When spring arrives, sunlight returns to the Antarctic and the active chlorine-containing species which have been released are then photolysed to produce $Cl^\bullet$ and $ClO^\bullet$ which are then able to destroy ozone.

(vi) Ozone destruction. It is believed that because (unlike at other latitudes) there is little atomic oxygen present in the polar stratosphere, ClO provides the key route to ozone destruction as set out in the following scheme:

$$ClO + ClO + M \rightarrow Cl_2O_2 + M$$
$$Cl_2O_2 + h\nu \rightarrow Cl + ClOO^\bullet$$
$$ClOO^\bullet + M \rightarrow Cl + O_2 + M \qquad (4.53)$$
$$2 \times (Cl + O_3 \rightarrow ClO + O_2)$$
$$\text{Overall}: 2O_3 + h\nu \rightarrow 3O_2$$

Once the spring is well advanced the sun breaks down the vortex and the air is mixed with that at lower altitudes, the PSCs evaporate and the atmosphere returns to 'normal'. During spring and summer ozone levels rise again, but the results suggest that they no longer return to the original pre-1980 values.

The ozone hole is, therefore, formed as a consequence of a critical relationship between the dynamics, the photochemistry and the surface chemistry of the system.

4.7.2 Reactions on polar stratospheric clouds

Most of the adsorption systems and reactions we have considered in this chapter and in Chapter 3 have been concerned with adsorption and reaction on the surface with desorption of a product, leaving the surface largely unchanged. For heterogeneous reactions in the atmosphere it is common for the surface to behave as one of the reactants; that is if a product is desorbed, it often contains atoms or molecules which were originally part of the surface. This is because the nature of the surface is rather different in atmospheric reactions compared with those involved with catalysts, in that the surface itself reacts to form volatile products.

While most atmospheric scientists are convinced that the outline of the chemistry involved in ozone depletion is as described in Section 4.7.1, far more detail is required to evaluate the importance of each of the elementary reactions involved. To this end the application of surface science techniques to model studies of reactions on ices, such

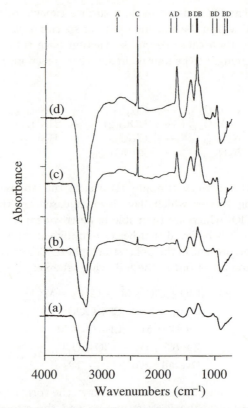

Fig. 4.16 The reaction of N_2O_5 with H_2O ice. (a) The IR spectrum of the water ice film. (b) Initial adsorption of N_2O_5 leads to the observation of features A and B which are due to H_3O^+ and solvated NO_3^-, respectively. From (b) through (c) to (d), features C and D grow in, and are assigned to NO_2^+ and molecular HNO_3, respectively. (From Horn *et al.* 1994.)

as those found in PSCs (which consist of water and nitric acid trihydrate, NAT), are now being carried out.

Generally, films of ice of the order of tens of nanometres thick and NAT are condensed on cryogenically cooled foils of polycrystalline materials such as gold or nickel. They can then be studied using the desired technique both before and after adsorption of an appropriate adsorbate. An example of one of a wide range of approaches that have been adopted for studying this sort of system is described here. Our example uses an adaptation of the RAIRS method to look at the adsorbate vibrations of the reservoir species N_2O_5, with the aim of identifying the reactions it undergoes on PSCs. The sensitivity requirements for these measurements are not as high as for RAIRS off a single crystal and so an angle of incidence of 75° is used instead of grazing angles.

The results for the deposition of N_2O_5 on a pure H_2O film at 140 K are shown in Fig. 4.16. The initial IR spectrum in Fig. 4.16(a) shows the vibrations of the H_2O molecules in the film itself, with the strongly hydrogen-bonded OH stretch in the 3300 cm^{-1} region and the HOH deformation in the 1600 cm^{-1} region. On initial adsorption of N_2O_5 (Fig. 4.16(b)), the features labelled A and B begin to grow in. The peaks labelled A are due to the formation of the hydronium ion (H_3O^+), while those labelled B are due to a solvated nitrate ion, NO_3^-. The observation of both these species together indicates that an amorphous layer of nitric acid hydrate, $2H_3O^+NO_3^- \cdot (n-3)H_2O$, has formed (where n depends on how many water molecules are available for solvation). These findings provide clear evidence that nitric acid is formed following the adsorption of the N_2O_5 reservoir species on a water ice cloud.

If further N_2O_5 is dosed onto the surface, the amount of water which is available in the surface layer to take part in the reaction is reduced and the features labelled C and D in Fig. 4.16(b), (c) and (d) grow in. Species C is assigned to a linear NO_2^+ species, while the peaks labelled D are due to molecular HNO_3. This is a very interesting observation, because it means first that subsurface water does not diffuse to the surface to take part in the reaction, as might be expected. Second, the initial reaction to form the hydrated nitric acid species takes place fairly readily. Once a certain proportion of the available surface water has been used up in the reaction, that reaction pathway is poisoned by further adsorption of N_2O_5. Rather than this inhibiting further reaction, the reaction mechanism changes to produce the molecular nitric acid species and the NO_2^+.

It was found that if the film shown in Fig. 4.16(d) was heated gently, the molecular form of the nitric acid was gradually converted to the ionic hydrate because water then became sufficiently mobile to diffuse from the bulk to the surface and interact.

4.8 Surface photochemistry

It would not be reasonable to leave the subject of reactions and reactivity on surfaces without mentioning a growing area of importance, that of surface photochemistry. The interest in this complex field arises because of photon-induced processes used in electronic and optoelectronic materials. Such processes include photon-induced metal deposition, etching, oxidation, nitridation and doping of semiconductor materials. Photocatalytic reactions have also been observed in photoelectrochemistry and on catalysts where, for example, photons have been used for splitting water on TiO_2 and in nitrogen reduction on Fe dispersed on a TiO_2 support. There are two likely mechanisms by which the photon-induced reaction can proceed.

4.8.1 Direct absorption of a photon by the adsorbate

Photodissociation has been observed for metal carbonyl compounds such as $Mo(CO)_6$. When physisorbed on metal or semiconductor substrates this can be photolysed to dissociate on the surface. The maximum dissociation yield occurs when the photolysis uses radiation with a wavelength of 290 nm, which corresponds to a metal-to-ligand charge transfer transition in the complex. In this case the maximum photodissociation yield is found to be independent of the substrate on which the carbonyl is adsorbed and this means that the mechanism proceeds by direct absorption of a photon by the adsorbate. The substrate plays only a minor role in the process.

4.8.2 Absorption of photons by the substrate

In photocatalytic and photoelectrochemical reactions, the driving force for reaction comes from the charge carriers (either electrons or holes) which are induced in the substrate by absorption of the incident photons. These charge carriers then interact with the adsorbates to induce the chemical reaction. The interaction can be such that the electron density in the adsorbate is rearranged to a certain extent and the adsorbate then undergoes a reaction to relax back to a more stable configuration. This sort of mechanism has been found for the photon-induced desorption of NO from Si and for the photon-induced dissociation of $Mo(CO)_6$ on potassium-modified copper. The potassium provides additional electron density to the electronic structure of the copper and so when photolysed the substrate can produce 'hot' electrons which attach themselves to the $Mo(CO)_6$. The negatively charged $Mo(CO)_6$ species then dissociates in order to relax. The whole process is called dissociative electron attachment.

Excitation by either mechanism is followed by desorption, dissociation, or reaction between co-adsorbed species. In order to further investigate the mechanism of the surface-photochemical reaction it is vital to study the photon yield as a function of incident wavelength, as this provides the key for identifying the nature of the electron transfer which initiates the reactions. In the future, direct excitation of a specific bond by the use of femtosecond or faster lasers might enable us to directly probe the dynamics of surface photochemistry, as in principle the excitations operate at similar rates to processes such as bond formation and cleavage.

5
Ultrathin films and interfaces

The most important consideration when discussing catalytic processes is to look at situations where the substrate–adsorbate interactions dominate with respect to adsorbate–adsorbate interactions. In the previous chapters we have looked at chemisorption and surface reactions where this is the case. It might be argued that if it were not for these adsorbate–adsorbate interactions, surface reactions would not proceed. While this is essentially true, it is important to realize that the key to the reaction happening at all lies in the fact that adsorption has occurred first.

In this chapter we will discuss ultrathin films and the interfaces between two different materials. In this case the adsorbate–adsorbate interactions are often just as or even more important than the substrate–adsorbate interactions. The driving force behind work in this area is the application of surface science in the microelectronics industry. Electronic devices consist of (typically) a semiconductor wafer on which there are layers of different materials that have been partially etched away. The interfaces act as junctions of different types, depending on the materials involved. An example of the sort of structure used was given in Chapter 3, Section 3.6.1.4.

In recent years it has been found that if a complete layer of one material is grown on another (e.g. one metal on another), the competition between adsorbate–adsorbate interaction and substrate–adsorbate interaction leads to some very interesting properties for the overlayer film that are quite different to those observed for the bulk material. These different properties can be electronic, structural, chemical and/or magnetic and may be significant in terms of catalysis.

5.1 Growth of ultrathin films

How ultrathin films grow on surfaces will be considered here in the context of the growth of atoms on a single-crystal substrate and of the formation of layers of molecules physisorbing on the surface. The models are equally applicable in both situations.

5.1.1 Growth modes

As atoms (or physisorbed molecules) accumulate on the surface, they do not necessarily form a single monolayer. Whether or not they do this depends on how much the adsorbate 'wets' the surface. In other words, is the energy gained as a result of bonding to the surface sufficient to overcome the energy required to form a large area of adsorbate which is only a single atom thick? It may be that it is energetically more favourable for the adsorbate to form 'clumps' more correctly called 3-d crystallites on the surface.

So the situation can be thought of as the competition between the adsorption energy and the surface tension of the adsorbate. Even if it is favourable for a monolayer to form, it is found that the surface tension plays an important role. This is because there is a tendency, as the first monolayer grows in, for the adsorbates to arrange in 2-d islands or patches which have the density expected for the full monolayer, surrounded by a much less densely packed, 2-d gas of adsorbate atoms.

Growth of an adsorbed layer past a single monolayer changes the nature of the surface on which further atoms or molecules adsorb, and so often changes the adsorption energy. This means that it is possible, following completion of each layer, that the growth mode will change from wetting of the substrate to forming bulk-solid crystallites. Formally, we can describe the growth of an ultrathin film in terms of four possible models which are shown in Fig. 5.1.

(i) Layer-by-layer growth (Frank–Van der Merwe). This is shown in Fig. 5.1(a) and describes the situation for 'ideal' growth of a film. The important feature is that as the atoms are deposited on the surface, they form layers. Each layer is completed before the next one starts to grow.

(ii) Stranski–Krastanov (SK) growth. This is shown in Fig. 5.1(b). Here, the initial layers grow in layer-by-layer, as in (i) but at some point the ordered sequence where complete layers form gives way to the formation of bulk crystallites.

(iii) Volmer–Weber (VW) growth. In this formation process, the adsorbates do not 'wet' the surface and so form bulk crystallites on the surface as shown in Fig. 5.1(c).

In order to form a single monolayer on the surface, the adsorbing molecule must migrate from the point at which it landed on the surface to an island edge. Having reached the island edge it will then adsorb and increase the size of the island. For each of the growth modes (i)–(iii) it is assumed that the mobility of the adsorbates on the surface is sufficient that they can diffuse to the island edges. This is not always the case as it may be that the substrate temperature is too low to allow diffusion to take place. In this case the fourth growth mode results.

(iv) Simultaneous multilayer formation. This is shown in Fig. 5.1(d) and occurs when the surface dynamics are such that diffusion across the surface is too slow to enable the adsorbates to form large-scale ordered structures. It can be likened to the effect of throwing a snowball at a wall. The adsorbates (like snowballs) hit and stick where they initially land. This often results in an adsorbate layer with varying thickness. If the reason for the formation of simultaneous multilayers is that the substrate temperature is too low for diffusion to occur, it is found that warming enables the ordered phase to form.

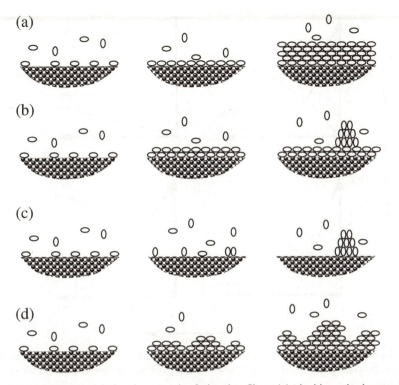

Fig. 5.1 The four models for the growth of ultrathin films: (a) ideal layer-by-layer growth, (b) Stranski–Krastinov, (c) Volmer–Weber, island formation, and (d) simultaneous multilayer formation. The ellipses represent molecules in the gas phase and adsorbed on the surface.

5.1.2 Monitoring ultrathin film growth

Having described the modes by which growth occurs, it is important that we are able to study them, as insights into the growth mode will provide us with valuable information on the energetics involved. Several methods are currently used to monitor growth and the most widely used are discussed here.

5.1.2.1 *Auger uptake curves*

Auger electron spectroscopy (Chapter 3, Section 3.6.1) finds an application in this field, in addition to all the others to which has been usefully applied. If we consider the inelastic mean free path (Chapter 2, Section. 2.4.3) for Auger electrons, it is clear that as a layer is covered by a new layer of adsorbate, the AES signal for the initial layer is attenuated. This means that if the intensities of an AES peak due to the substrate and an AES peak due to the adsorbate are monitored as a function of adsorbate deposition time, the shapes of the curves will be indicative of the growth mode. The expected curves for each of the growth modes (i)–(iv) are shown in Fig. 5.2.

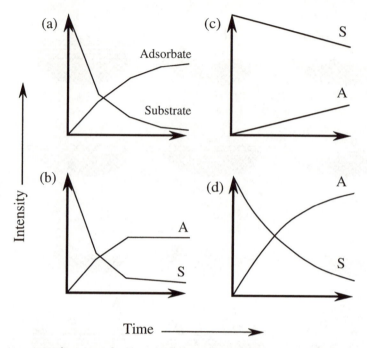

Fig. 5.2 Auger uptake curves for the adsorbate A and substrate S for the four growth modes: (a) ideal layer-by-layer growth, (b) Stranski–Krastinov, (c) Volmer–Weber, and (d) simultaneous multilayer formation.

(i) The AES uptake curves for layer-by-layer growth are shown in Fig. 5.2(a). As the initial layer of adsorbate grows in, the adsorbate and substrate intensities increase and decrease, respectively, with a steady slope, which arises because the intensity of signal is proportional to the amount in the surface layer. On formation of the complete first layer, there is a 'break' point where the gradients change to give a lower slope. The lower slope is found because, considering the mean free path, there is now less likelihood of (for example) an Auger electron from the substrate escaping from the surface. Break points occur each time a new layer is completed and once the adsorbate is thick enough, the signal due to the substrate disappears completely (usually at about six monolayers).

(ii) For VW growth, which is shown in Fig. 5.2(c), the adsorbate and substrate signals increase/decrease steadily as the crystallites form. There can be crystallites containing many adsorbate layers, in coexistence with patches of the bare substrate, which means that there will be no 'break points' in the AES curves.

(iii) For SK growth, the AES curves are initially the same as for layer-by-layer growth as shown in Fig. 5.2(b), but when crystallites begin to form, the uptake curves begin to resemble those of VW growth.

(iv) For simultaneous multilayer formation, smooth curves are found as the adsorbate signal increases and the substrate signal decays with the random deposition process, as shown in Fig. 5.2(d).

There are a few drawbacks when using AES for monitoring film growth. For example, the AES peak energy and shape may change if the chemical environment changes. The detection efficiency of the analyser may be different for the adsorbate and substrate Auger electron energies and is also affected by diffraction of the incident and Auger electrons. The inelastic mean free path depends on the density of valence electrons and so this may vary depending on how different the valence electron densities are for adsorbate and substrate. Finally, the cross-sections for ionization can change with surface composition. These factors can sometimes make it difficult to determine whether or not, and where, break points occur in the uptake curves.

Generally, however, determining Auger uptake curves has proved to be of considerable value in the study of the growth of ultrathin films.

5.1.2.2 Diffraction methods

These methods rely on the interference of diffracted beams, which are scattered from the surface.

(i) Reflection high-energy electron diffraction (RHEED)

The most commonly used method is reflection high-energy electron diffraction (RHEED). This can be used to monitor the film as it grows, and works by firing a beam of high-energy electrons ($\sim$ 5–50 keV in energy) at grazing angles of incidence to the surface. The specularly scattered beam intensity is then monitored as a function of deposition time. A typical result for the RHEED intensity variation for layer-by-layer growth is shown in Fig. 5.3 for the growth of Ni on W(110). The results show striking oscillations and the maximum intensity peaks of these oscillations coincide with the completion of each layer.

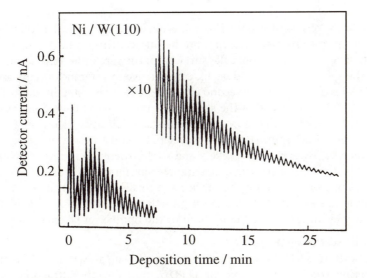

Fig. 5.3 RHEED oscillations for layer-by-layer growth of a metal-on-metal system, in this case Ni growing on W(110). (From Koziol *et al.* 1987.)

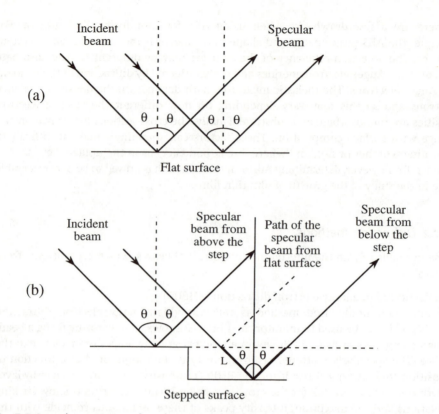

Fig. 5.4 Specular scattering of an electron or He beam from (a) a flat surface and (b) a stepped surface.

The reason for the observation of oscillations is obvious if one considers that they are the result of Bragg scattering (eqn (2.33)) from the surface. This is illustrated in Fig. 5.4, in which scattering takes place from a flat surface. For the zero-order diffraction, that is detection of the specularly scattered beam, constructive interference occurs and so a maximum signal is detected. In the second case, where there is a step in the surface, as shown in Fig. 5.4(b), it is clear that the specular beam scattered from below the step travels further than that scattered from above the step (distance $2L$ on the figure). If this distance is equal to an exact integer wavelength, constructive interference will occur; however, this is extremely unlikely and so destructive interference is usually found. Obviously, the largest amount of destructive interference is found when there is the maximum number of steps on the surface and this is most likely at $\theta = 0.5$.

No oscillations are expected for the other growth modes and so RHEED is used to distinguish layer-by-layer growth and to monitor the number of layers grown.

(ii) Helium atom scattering (HAS)
Helium atom scattering (HAS) can also be used to investigate growth modes because of similar scattering considerations. As for LEED patterns, with diffraction patterns obtained from HAS it is found that a higher degree of order is required for acquiring

useful data, but once obtained the interpretation is less complicated because kinematic rather than dynamical scattering models can be used. With HAS it is possible to study the oscillations as a function of scattering angle and from the results, quantitative information on the growth mode can be obtained. It is also possible to use these results to distinguish between all four of the accepted growth modes.

5.2 Orientation and strain

It is interesting to turn our attention to the structures of the adsorbate layers which result from the above growth processes. Once one material is brought into contact with another, there will be energy differences which will influence the structure and orientation of the adsorbate layers. These effects may be such that the overlayer does not fall, or only partially falls, into registry with the substrate structure. If it does fall into registry, the overlayer formed may be 'strained' and it attempts to alleviate the strain by expanding or contracting vertically. In this section we look at these effects in detail.

5.2.1 Commensurate and incommensurate structures

When atoms and molecules adsorb at low coverages there is a tendency for them to form a so-called 'commensurate' structure with the underlying substrate. This means that the overlayer has a (1×1) structure with the substrate acting as a template. The formation of such overlayer films is called *epitaxial growth*. In some extreme cases, as the overlayer grows, the substrate imposes its own crystal structure, orientation and lattice parameter on the overlayer. In this case the growth process is usually termed *pseudomorphic growth*.

When adsorbates form a condensed layer, the adsorbate–adsorbate and the substrate–adsorbate forces control the structure of the layer formed. The adsorbate–adsorbate forces attempt to impose a structure on the layer which is similar to that of the bulk solid of the adsorbate. This might be very different from the structure of a commensurate layer on the substrate, and the structure adopted depends on which of the two types of forces is strongest. Comparatively strong adsorbate–adsorbate forces will produce a layer not in registry with the surface, and these are known as an 'incommensurate' layers.

In many cases the adsorbate–adsorbate forces are of very similar strength to the substrate–adsorbate forces and as neither are particularly dominant, the structure of the adsorbate layer will have a very sensitive dependence on the adsorbate density in the layer and/or on the temperature of the system. The balance of forces produces complex phase diagrams for some surface–adsorbate systems. During adsorption of a monolayer, the overlayer might undergo a transition from a commensurate to an incommensurate phase because of the increase in adsorbate density. This might not occur simultaneously across the surface, and patches (or domains) of commensurate

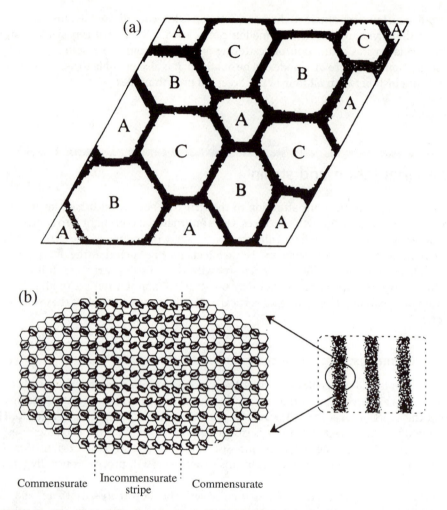

Fig. 5.5 The formation of incommensurate domain walls in physisorbed overlayers on graphite. (a) Domain walls in adsorbed Kr attract each other leading to the hexagonal domain walls which are observed. (From Kock *et al.* 1984.) (b) Parallel domain walls result from the repulsive interactions of adsorbed N_2. (From Wilkes 1990.)

phase will remain with incommensurate domain walls (made up of extra atoms) forming around them. The domain walls interact with each other and the structure resulting depends on whether their interaction is attractive or repulsive. An example of these two possibilities is shown in Fig. 5.5. Figure 5.5(a) shows the effect of domain walls attracting each other which happens for Kr on graphite where hexagonal domains are formed. For a repulsive interaction, such as is seen in Fig. 5.5(b) for molecular nitrogen on graphite, parallel stripes are formed by the domain walls. In either case the addition of more adsorbate atoms/molecules results in an increase in the domain walls in both size and number, until the entire structure becomes incommensurate.

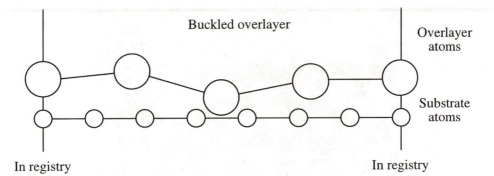

Fig. 5.6 'Buckling' of an incommensurate overlayer in order to fit a larger-scale commensurate layer.

Even in a totally incommensurate layer, the surface continues to affect the overlayer structure. It might impose a much larger commensurate unit cell on the structure. An indication of how this could happen is shown in Fig. 5.6, where every fifth overlayer atom sits directly above a substrate atom. This registry occurs across the distance occupied by eight substrate atoms and some 'buckling' of the overlayer is needed for it to fit across that distance. As the structure shown in Fig. 5.6 illustrates, the fact that some of the overlayer atoms are in registry with the surface and some are not means that it sometimes becomes difficult to make a distinction between a commensurate and an incommensurate structure.

5.2.2 Epitaxial strained layers

Epitaxy occurs when the overlayer material 'distorts' from its bulk lattice to fit in registry with the surface. This is often called 'row-matching' and the distortion occurs to cope with the 'misfit' of the two lattices. A useful parameter is the misfit f which is given by

$$f = \frac{(a - b)}{a} \tag{5.1}$$

where a and b are the nearest-neighbour distances of the two materials. For small values of the misfit, the overlayer often simply compresses or expands to compensate, leaving it in a situation of some strain. It might be possible under these conditions for an overlayer to adopt a crystal structure different from that normally found in the bulk. For example, when Fe, which is normally body-centred cubic, is grown on Cu(100), it initially adopts the face-centred cubic structure of the underlying copper lattice. It turns out that there is a stable phase (above 1100 K) of bulk Fe, which has an fcc structure. In this structure its lattice parameter is very close to that of the copper and so when deposited on the copper surface it adopts a metastable fcc structure at a much lower temperature.

In cases of a slightly larger misfit, the overlayer might rotate to give the lowest possible overall energy for the system. To attain the lowest energy it is necessary for the maximum number of individual overlayer atoms to sit at the energy minimum of the

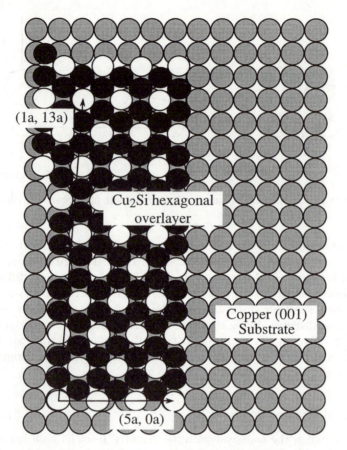

(1a, 13a)

Cu₂Si hexagonal
overlayer

Copper (001)
Substrate

(5a, 0a)

Fig. 5.7 The real space structure of a SiCu₂ overlayer on a Cu(100) surface, determined from HAS. The overlayer is rotated slightly as shown in order to achieve the lowest possible surface energy. (From Graham *et al.* 1994.)

substrate, that is in the sites that atoms in a new layer of substrate material would sit. For the rotated overlayer, the overlayer atoms do not line up with the substrate but do manage to obtain the lowest total energy on average across the surface. The overlayer adopted is ultimately determined by the competition between the strain energy (due to compression/expansion required by going into registry) and the minimum total energy that can be obtained by rotation of the overlayer.

An example of this type of rotation is shown in Fig. 5.7 for Si deposited on Cu(100). An overlayer consisting of a stoichiometry of 1:2 of Si:Cu atoms is formed on top of the Cu lattice. The overlayer takes up a hexagonal structure. A periodicity of 5a is required in one direction and 13a in the other, rotated by 4.4° with respect to the copper substrate, is required to fit the overlayer onto the substrate, as shown. (Black circles indicate the Cu overlayer atoms and white circles the Si overlayer atoms.) In this case the 5a direction is not rotated (so the overlayer is effectively 'sheared' with respect to the substrate).

5.3 Properties of ultrathin films

The effect of growing one material on another has interesting implications for the properties of the resulting material. It might be that an alloy forms which has a structure different from that of the original materials. At the other extreme, complete non-wetting of one material by the other, such as is seen for Al grown on Si, produces novel 'Schottky barriers' which are useful in electronic devices.

In Section 5.2 we considered the interactions of adsorbates and substrate, and how this affected the resulting structure. The nature of the resulting overlayer film is dependent on how the adsorbate interacts with the substrate and other adsorbates. Because of the interplay of the forces, these overlayer films often exhibit unusual properties and the effects observed are described below.

5.3.1 Electronic and chemical properties

The electronic structure of the surface of an ultrathin film is usually different from the electronic structure of the same element where it terminates the bulk material. In the case of alloy formation it is obvious that there will be substantial mixing of the electronic states of the adsorbate with the substrate. In the case of a 'pure' film which grows on the substrate, the electronic state (which can be thought of in terms of the density of states and electronic structure) is affected because of an interaction with the electronic states of the underlying substrate. It is also affected if a different interatomic distance is adopted by the overlayer atoms (than is usually found for the bulk of the same material), which has the effect of either increasing or reducing the density of the electrons in the film. This alteration to the density of states has considerable influence on the chemistry of the films, and so adsorption and reactivity are often significantly different on the ultrathin film when compared to the bulk crystal.

5.3.1.1 The Rh(100)/Fe/H$_2$ system

This effect is particularly noticeable for ultrathin films which have different structures compared to those of the bulk material, like Fe/Cu(100). Another system where a structural change is made is for the growth of iron on the Rh(100) substrate. Like the Fe/Cu system, the iron takes up an fcc structure for the first few layers. PES measurements (see Fig. 5.8(a)) show that there are two surface states due to the d-bands at the surface, for the clean Rh surface. These are at binding energies of 0.8 and 2 eV. Deposition of a monolayer of Fe shifts both of these surface states by about 0.4 eV to larger binding energy as shown in Fig. 5.8(b); with little effect on the PES peaks arising from the bulk. It was also found that there was a large reduction in density of states at the Fermi level. As additional layers (2 and 3 monolayers) of Fe were deposited it was found that the density of states at the Fermi level increased again. The rationalization of these observations is that in the initial monolayer of Fe, the d-bands of the Fe hybridize with the d-bands of the Rh, so that the bonding state is shifted down in energy (to higher binding energy) and the antibonding state is pushed up in energy. For thicker two and three monolayers films there is less hybridization and so the density of states increases again.

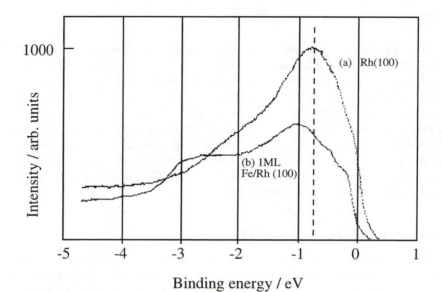

Fig. 5.8 UPES of clean Rh(100) and 1 ML Fe on Rh(100), measured normal to the sample. (From Egawa *et al.* 1994.)

This change in density of states should affect the adsorption properties of the surface. Fig. 5.9 shows the temperature programmed desorption (see Chapter 3, Section 3.4.2) of hydrogen from the clean and Fe-covered Rh surface.

In our example, the surfaces were each exposed to six Langmuirs of hydrogen and the coverages resulting are given (these are obtained by finding the integrated intensity of the TPD peaks). From these intensities it is clear that hydrogen adsorption on the bare Rh surface is far more favourable than on the Fe films. Additionally, the sticking coefficient decreases with Fe coverage. It should also be noted that the sticking coefficient for H_2 on the ultrathin Fe films is about ten times *larger* than that of H_2 on bcc surfaces.

The second feature of the TPD spectra is that desorption from the clean Rh(100) surface gives two peaks at 140 and 320 K. For a surface which has one monolayer of Fe deposited on it there is a single hydrogen desorption peak at 240 K. As the thickness of the Fe films is increased, this desorption temperature shifts back up to 300 K.

From the TPD the activation energy for desorption for the one-monolayer Fe film is found to be 61 kJ mol^{-1}; the bond energy for Fe–H is 245 kJ mol^{-1}. This is some 20 kJ mol^{-1} less than that for the Fe–H bond energy on the bcc Fe(100) surface, where the hydrogen is found to desorb above 400 K. The increase in desorption temperature (and thus of bond energy) with increasing Fe thickness is due to the interaction of the hydrogen with the increased density of states at the Fermi level.

5.3.2 Magnetic properties

Some materials display magnetic properties. Not too surprisingly, ultrathin films of them are found to exhibit a wide range of different magnetic effects arising from all of

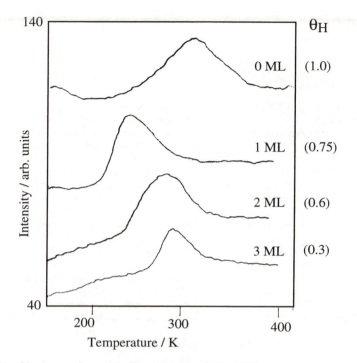

Fig. 5.9 TPD of hydrogen from thin films of Fe on Rh(100) following exposure to 6 L of hydrogen. The coverage of hydrogen is given in brackets. The Fe thicknesses are (a) 0 ML, (b) 1 ML, (c) 2 ML and (d) 3 ML. (From Egawa *et al.* 1994.)

the considerations which apply to the electronic structures described in Section 5.3.1. Regrettably, although this is a fascinating field of immense interest because of the potential applications in the fields of data storage media and optical communications, it is not usually considered under the umbrella of surface chemistry and is usually left to the surface physicists. It is notable, however, that the effect of magnetism on the chemistry of these films has received little attention but shows considerable potential for the future.

BIBLIOGRAPHY

GENERAL REFERENCES

Attard, G. and Barnes, C. (1998). *Surfaces*. Oxford Chemistry Primer no. 59, Oxford University Press.

Banwell, C. N. and McCash, E. M. (1994). *Fundamentals of molecular spectroscopy* (4th edn). McGraw-Hill.

Bond, G. C. (1974). *Heterogeneous catalysis: principles and applications*. Clarendon Press, Oxford.

Briggs, D and Seah, M. P. (1990). *Practical surface analysis. Vol. 1: Auger and X-ray photoelectron spectroscopy*. Wiley.

Briggs, D and Seah, M. P. (1992). *Practical surface analysis. Vol. 2: Ion and neutral spectroscopy*. Wiley.

Clark, R. J. H. and Hester, R. E. (1988). *Advances in spectroscopy. Vol. 16: Spectroscopy of surfaces*. Wiley.

Clark, R. J. H. and Hester, R. E. (1998). *Advances in Spectroscopy. Vol. 26: Spectroscopy for surface science*. Wiley.

Gasser, R. P. H. (1985). *An introduction to chemisorption and catalysis by metals*. Oxford Science Publications.

Ibach, H and Mills, D. L. (1982). *Electron energy loss spectroscopy and surface vibrations*. Academic Press.

Kittel, C. (1986). *Introduction to solid state physics*. Wiley.

Prutton, M. (1994). *Introduction to surface physics* (2nd edn). Oxford Science Publications.

Somorjai, G. A. (1994). *Introduction to surface chemistry and catalysis*. Wiley.

Thompson, T., Baker, M. D., Christie, A. and Tyson, J. F. (1985). *Auger electron spectroscopy*. Wiley.

Woodruff, D. P. and Delchar, T. A. (1988). *Modern techniques of surface science*. Cambridge University Press.

Zangwill, A. (1988). *Surface physics*. Cambridge University Press.

ACKNOWLEDGEMENTS FOR FIGURES

CHAPTER 2

Fig. 2.16 Redrawn from the *Handbook of X-ray photoelectron spectroscopy*, Perkin-Elmer Corporation (1992), with permission.

Fig. 2.20 Redrawn from *Chemistry in two-dimensions: surfaces*, G. A. Somorjai (1981), with kind permission from the author and Cornell University Press.

CHAPTER 3

Fig. 3.10 (a) Reprinted from *Journal of Physics and Chemistry of Solids*, **3**, 95–101, P. Kisliuk, The sticking probabilities of gases chemisorbed on the surfaces of solids (1957), with kind permission from Elsevier Science Ltd, The Boulevard, Langford Lane, Kidlington, OX5 1GB, UK.

Fig. 3.10 (b) Redrawn from *Surface Science*, **53**, 55–73, S. P. Singh-Boparai, M. Bowker and D. A. King, Crystallographic anisotropy in chemisorption: nitrogen on tungsten single crystal planes (1975), with kind permission from Elsevier Science—NL, Sara Burgerhartstraat 25, 1055 KV Amsterdam, The Netherlands.

Fig. 3.16 Redrawn from *Surface Science*, **283**, 427–37, A. Al-Sarraf, J. T. Suckless, C. E. Wartnaby and D. A. King, Adsorption microcalorimetry and sticking probabilities on metal single crystal surfaces (1993), with kind permission from the author and Elsevier Science—NL, Sara Burgerhartstraat 25, 1055 KV Amsterdam, The Netherlands.

Fig. 3.27 Redrawn from *Journal of the Chemical Society, Faraday Transactions*, **91**, 3563–7, J. P. Camplin, J. C. Cook and E. M. McCash, Reflection–absorption infrared spectroscopy at cryogenic temperatures (1995), with kind permission from the Royal Society of Chemistry.

Fig. 3.32 Redrawn from *Vacuum* **40**, 423–7, E. M. McCash, Surfaces and vibrations (1990), with kind permission from Elsevier Science Ltd, The Boulevard, Langford Lane, Kidlington, OX5 1GB, UK.

Fig. 3.34 Redrawn from *Journal of the Chemical Society, Faraday Transactions*, **86**, 2757–63, M. A. Chesters, C. De La Cruz, P. Gardner, E. M. McCash, P. Pudney, G. Shahid and N. Sheppard, An infrared spectroscopic comparison of the chemisorbed species from ethene, propene, but-1-ene, and *cis*- and *trans*-but-2-ene on Pt(111) and on a platinum/silica catalyst (1990), with kind permission of the Royal Society of Chemistry.

Fig. 3.37 (b) Redrawn from *Surface Science*, **139**, 87–97, B. J. Bandy, M. A. Chesters, M. E. Pemble, G. S. McDougall and N. Sheppard, Low temperature electron energy loss spectra of acetylene chemisorbed on metal single-crystal surfaces: Cu(111), Ni(110) and Pd(110) (1984), with kind permission from Elsevier Science—NL, Sara Burgerhartstraat 25, 1055 KV Amsterdam, The Netherlands.

Fig. 3.43 Redrawn from *The handbook of Auger electron spectra*, P. W. Palmberg, G. E. Riach, R. E. Weber and N. C. MacDonald (1972), with kind permission from the Physical Electronics Industries, Inc.

Fig. 3.45 (a,b) Redrawn from *Surface Science* **111**, 441–51, N. D. S. Canning, M. D. Baker and M. A. Chesters, Ethylene and acetylene adsorption on Cu(111) and Pt(111) studied by Auger spectroscopy (1981), with kind permission from Elsevier Science—NL, Sara Burgerhartstraat 25, 1055 KV Amsterdam, The Netherlands.

Fig. 3.45 (c) Redrawn from *Surface Science* **111**, 452–60, M. D. Baker, N. D. S. Canning and M. A. Chesters, Auger spectra of carbon monoxide chemisorbed on Pt(111) and Cu(111) (1981), with kind permission from Elsevier Science—NL, Sara Burgerhartstraat 25, 1055 KV Amsterdam, The Netherlands.

Fig. 3.45 (d) Redrawn from *Journal of Chemical Physics*, **55**, 2317–36, W. E. Moddeman, T. A. Carlson, M. O. Krause, B. P. Pullen, W. E. Bull and G. K. Schweiter, Determination of the K-LL Auger spectra of N_2, O_2, CO, NO, H_2O and CO_2. (1971), with kind permission from the authors and the American Institute of Physics.

Fig. 3.47 Reproduced from *Introduction to surface physics*, 2nd edn, M. Prutton, Oxford University Press (1994), with kind permission from the author and Oxford University Press.

Fig. 3.48 Redrawn from *Vacuum*, **36**, 1005–10, D. Briggs and M. J. Hearn, Interaction of ion beams with polymers, with particular reference to SIMS (1986), with kind permission from Elsevier Science Ltd, The Boulevard, Langford Lane, Kidlington, OX5 1GB, UK.

CHAPTER 4

Fig. 4.8 Redrawn with permission from *Journal of Physical Chemistry*, **99**, 13755–8, K. Wilson, C. Hardacre and R. M. Lambert, SO_2-promoted chemisorption and oxidation of propane over Pt(111). Copyright 1995, the American Chemical Society.

Fig. 4.9 Redrawn with permission from *ACS Symposium Series*, **638**, 394, K. Wilson, C. Hardacre and R. M. Lambert, SO_2-promoted propane oxidation over Pt(111). Copyright 1996, the American Chemical Society.

Fig. 4.10 Redrawn from *Surface Science*, **15**, 443–65, G. Ertl and P. Rau, Chemisorption und katalytische von sauerstoff und kohlenmonoxid an einer palladium (110)-oberfläche (1969), with kind permission from Elsevier Science—NL, Sara Burgerhartstraat 25, 1055 KV Amsterdam, The Netherlands.

Fig. 4.11 Redrawn with permission from *Physical Review Letters*, **49**, 177–80, G. Ertl, P. R. Norton and J. Rüstig, Kinetic oscillations in the platinum-catalysed oxidation of CO. Copyright 1982, the American Physical Society.

Fig. 4.12 Redrawn from *Journal of Chemical Physics*, **98**(12), 9977–85, S. Nettlesheim, A. von Oertzen and G. Ertl, Reaction diffusion patterns in the catalytic CO-oxidation on Pt: front propagation and spiral waves (1993), with kind permission from the author and the American Institute of Physics.

Fig. 4.14 Redrawn from *Surface Science*, **357–8**, 602–8, C. T. Rettner and D. J. Auerbach, Dynamics of the formation of HD from D(H) atoms colliding with H(D)/Cu(111): a model study of an Eley–Rideal reaction (1996), with kind permission from Elsevier Science—NL, Sara Burgerhartstraat 25, 1055 KV Amsterdam, The Netherlands.

Fig. 4.16 Reprinted with permission from *Journal of Physical Chemistry*, **98**, 946, A. B. Horn, T. Koch, M. A. Chesters, M. R. S. McCoustra and J. R. Sodeau, A low temperature study of the reactions of the stratospheric NOy reservoir species, dinitrogen pentoxide with water ice, 80–160 K. Copyright 1994, the American Chemical Society.

CHAPTER 5

Fig. 5.3 Redrawn from *Applied Physics Letters*, **51**, 901, C. Koziol, G. Lillienkamp and E. Bauer (1987), with kind permission from the authors and the American Institute of Physics.

Fig. 5.5 (a) Redrawn from *Surface Science*, **145**, 329, S. W. Kock, W. E. Rudge and F. Abraham, The commensurate transition of krypton on graphite: a study via computer simulation (1984), with kind permission from Elsevier Science—NL, Sara Burgerhartstraat 25, 1055 KV Amsterdam, The Netherlands.

Fig. 5.5 (b) Redrawn from J. L. Wilkes, PhD thesis, University of Cambridge (1990), with permission.

Fig. 5.7 Reprinted from *Physical Review B*, **50**(20), 15304, A. P. Graham, B. J. Hinch, G. P. Kochanski, E. M. McCash and W. Allison, Two-dimensional silicide (5×3) structure on Cu(100) as seen by scanning tunnelling microscopy and helium atom scattering (1994), with kind permission from the American Physical Society.

Figs 5.8 and 5.9 Redrawn from *Surface Science Letters*, **304**, L488–92, C. Egawa, S. Oki and Y. Murata, Electronic structure and reactivity of ultra-thin Fe films on a Rh(100) surface (1994), with kind permission from Elsevier Science—NL, Sara Burgerhartstraat 25, 1055 KV Amsterdam, The Netherlands.

Index